Measuring Wellbeing
Towards sustainability?

Karen Scott

Routledge
Taylor & Francis Group

LONDON AND NEW YORK

First published 2012
by Routledge
2 Park Square, Milton Park, Abingdon, Oxon OX14 4RN

Simultaneously published in the USA and Canada
by Routledge
711 Third Avenue, New York, NY 10017

Routledge is an imprint of the Taylor & Francis Group, an informa business

British Library Cataloguing in Publication Data
A catalogue record for this book is available from the British Library

Library of Congress Cataloging-in-Publication Data
Scott, Karen.
Measuring wellbeing : towards sustainability? / Karen Scott.
    p. cm.
  Includes bibliographical references and index.
  1. Well-being. 2. Quality of life. 3. Social indicators. 4. Sustainable
development. I. Title.
  HN25.S384 2012
  306–dc23   2012001471

ISBN: 978-1-84971-462-4 (hbk)
ISBN: 978-1-84971-463-1 (pbk)
ISBN: 978-0-203-11362-2 (ebk)

Typeset in Times New Roman
by HWA Text and Data Management, London

Printed and bound in Great Britain by
TJ International Ltd, Padstow, Cornwall

Measuring Wellbeing

Improving wellbeing and sustainability are central goals of government, but are they in conflict? This engaging new book reviews that question and its implications for public policy through a focus on indicators.

It highlights tensions on the one hand between various constructs of wellbeing and sustainable development, and on the other between current individual and societal notions of wellbeing. It recommends a clearer conceptual framework for policy makers regarding different wellbeing constructs which would facilitate more transparent discussions. Arguing against a win-win scenario of wellbeing and sustainability, it advocates an approach based on recognising and valuing conflicting views where notions of participation and power are central to discussions.

Measuring Wellbeing is divided into two parts. The first part provides a critical review of the field, drawing widely on international research but contextualised within recent UK wellbeing policy discourses. The second part embeds the theory in a case study based on the author's own experience of trying to develop quality of life indicators within a local authority, against the backdrop of increasing national policy interest in measuring 'happiness'.

This accessible and informative book, covering uniquely both practice and theory, will be of great appeal to students, academics and policy makers interested in wellbeing, sustainable development, indicators, public policy, community participation, power and discourse.

Dr Karen Scott is a research fellow at The Centre for Rural Economy at Newcastle University, UK. Her professional experience in social care, community planning and environmental fields underpins her academic research interest in the politics of human wellbeing and sustainable development at the local level.

Contents

Illustrations

Figures

Tables

Boxes

Preface

This book started in 2004 when I naively began what seemed like a straightforward three-year project. I joined a local authority strategy team to help create a set of quality of life indicators. The plan was that through consultation with policymakers and residents, a local vision of wellbeing would emerge which could be turned into a set of measurements to inform policy. During this time a new wave of interest in happiness emerged at the national level, contributing to existing conflicts between policymakers in the local authority over whether indicators should reflect sustainability or subjective wellbeing. These conflicts were linked not only to different interpretations of wellbeing but to different ideas of participation and policymaking. The task of developing indicators proved so difficult and contested that measures were not produced. What did happen, however, was something which I believe was far more valuable – a small but significant shift in the dominant policy culture towards a more rigorous scrutiny of how (and whose) wellbeing would be improved by existing economic growth agendas.

Although much literature discusses measuring wellbeing, quality of life or sustainability at the local level, this tends to focus on community 'how to' guides or best practice case studies. This literature was inadequate in helping me to understand my experience as it rarely included a critical and detailed appraisal of the realities of policymaking and the role of power, culture and discourse. So, throughout this book I will be questioning the values and strategic agendas that may underpin various discourses of wellbeing and its measurement, including a critique of the notion of measurement. Although the book is contextualised within UK national policy discourses on wellbeing, it is written with local governance in mind. In the post-2008 financial climate, as more and more emphasis is put on localism, the role of local government and community participation in defining and measuring wellbeing is key. There are, of course, a variety of factors which vary to create unique situations in every local area, and wellbeing is a political concept which is heavily contextualised. More detailed ethnographic data is needed to understand how wellbeing discourses work in different areas and what effect they have. This book presents one such case study.

I have attempted to provide what would have been useful for me as I struggled to negotiate a complex and contested project. I offer a critical introduction to: the variously framed concept of wellbeing and its measurement; its relationship with sustainable development; the role of power and participation in the definition and measurement of wellbeing; and the role of knowledge and indicators in public policy. The second part of the book looks at these broad and interconnected fields through a detailed account of the project above, including the difficulties and conflicts which arose.

If you have picked up this book expecting a recipe for sustainable wellbeing, with a neat set of indicators to match, you will be disappointed. There is already a profusion of literature offering these sorts of prescriptions, some of which is critically reviewed in this book. However, if you are searching for an introductory text about measuring wellbeing which is grounded in the experience of our everyday lives and the realities of politics and policymaking, then read on.

Acknowledgements

I am grateful to many people for helping me produce this book. My thanks to:

Derek Bell, Tim Gray and William Maloney in the Department of Politics, Newcastle University and Yvonne Rydin at University College London.

Those at Blyth Valley who participated in and supported this research. I am sorry I cannot acknowledge each person by name, but I need to respect confidentiality.

Alison Kuznets, Charlotte Russell, Khanam Virjee and Joanna Endell-Cooper at Earthscan/Taylor and Francis and John Hodgson and Holly Knapp at HWA. Thank you also to the three anonymous reviewers whose comments were very useful. Thank you to Cambridge University Press for their kind permission to use copyrighted material from Martha Nussbaum's book *Women and Development*.

Wellbeing researchers in other institutions for many conversations which have informed my thinking, especially Sarah Atkinson, Durham University and Ian Bache, Sheffield University, and Jo Swaffield for her help in organising a successful wellbeing workshop at Newcastle University in 2010.

My colleagues in the Centre for Rural Economy at Newcastle University, a wonderfully supportive place to work. In particular, I am deeply indebted to Sue Bradley for her kindness, without which I doubt I would ever have finished this book. Thanks to Menelaos Gkartzios for the many coffee breaks.

Finally, my deep thanks to Gary Cordingley, for so much more support than I deserve and last but not least to my children Megan Scott and Bryn Scott for putting up with it all.

The deficiencies in the work presented here are entirely mine.

Acronyms

AC	Audit Commission
BVBC	Blyth Valley Borough Council
CCT	compulsory competitive tendering
CPA	comprehensive performance assessment
DCLG	Department of Communities and Local Government
Defra	Department for Environment, Food and Rural Affairs
DETR	Department of Environment, Transport and the Regions
ESRC	Economic and Social Research Council
GDP	gross domestic product
GNP	gross national product
GVA	gross value added
IMD	Index of Multiple Deprivation
LGA	Local Government Act
LGMA	local government modernisation agenda
LSP	local strategic partnership
NCC	Northumberland County Council
NDC	New Deal for Communities
NEF	New Economics Foundation
NPM	new public management
ODPM	Office of the Deputy Prime Minister
PAGE	Partnership Action Group Executive
QLIs	quality of life indicators
SD	sustainable development
SENNTRi	South East Northumberland and North Tyneside initiative
SRB	single regeneration budget
SWB	subjective wellbeing

Part I

Reviewing wellbeing concepts and measurement

1 Introduction

The death of economics and the rise of wellbeing

While I was writing this book, the UK Coalition Government decided to develop a national measure of wellbeing. Sceptics questioned the political wisdom of doing this during a fiscal squeeze as public sector cuts, rising unemployment and threats of trade union action began to hit home. However, early in 2011 the Office of National Statistics held face-to-face and online consultations throughout the UK. This collected a total of 34,000 responses (ONS 2011). At the session I attended, fifty of us had ninety minutes to give our views, not only on what wellbeing is, but also on how it should be measured. Our opinions were meticulously gathered up on flip chart sheets each covered with a rash of post-it stickers so that 'what matters' can inform measurements and indicators for national policy. The UK is not the only country embarking on this endeavour. The great and the good in international wellbeing measurement research are feverishly constructing wellbeing indices and publishing dense reports in a seemingly exponential curve of knowledge production. Australia, Canada, Denmark, Finland, France, Germany, Ireland, Italy, Norway, Spain and the UK are among the countries striving to develop national wellbeing accounts to rival Bhutan's famous Gross National Happiness.

The recent rise and rise of wellbeing measurement is linked to a new wave of dissatisfaction with the neoliberal economic project and its pervasive influence on our lives, our culture and the planet. It is now widely recognised, documented and evidenced that a narrow focus on economic growth is unsustainable and that GDP is an inappropriate indicator of social progress. The advances in relative wealth and quality of life that market democracies have enjoyed are compromising ecological systems in fundamental ways[1] while failing to reduce inequalities[2] or increase life satisfaction.[3] The increasing belief among many that 'we have got close to the end of what economic growth can do for us' means scholars and policymakers are looking more seriously at alternative conceptions of development in these countries (Wilkinson and Pickett 2010: 5). At the time of writing, during global financial instability, tented communities of activists are occupying many city centres worldwide in protest at wealth inequalities. While the 2008 financial crisis may have been the last nail in the coffin of political legitimacy for neoliberalism, it may continue to exist in an increasingly erratic 'zombie state' until an alternative development path is institutionalised (Peck 2010; Peck *et al.* 2009). Economist Manfred Max-Neef (2011: xviii) argues that neoliberalism represents the 'death of true economics' and calls for a return to Aristotle's original idea of 'oikonomia' which supports 'the art of living and living well'. In Aristotle's account of economics, acquisition of material wealth and money-making was subordinate to household prudence and civic virtue, the latter being strongly linked to his idea of human flourishing or 'eudaimonia'.[4] This idea of human flourishing has been taken up by a wide variety of policy actors across the globe, resulting in a surge of interest in 'redefining progress' and moving 'beyond GDP'. In the wake of the economic paradigm of progress which is 'at least severely wounded if not dead'[5] eudaimonic wellbeing is fast emerging as a new paradigm of development, alongside the now well-established theory of sustainable development, bringing with it a new industry of wellbeing measurement to challenge the dominance of the GDP indicator.

The UK has been at the forefront of recent work on wellbeing, with activities spearheaded by social reformist think-tanks such as the New Economics Foundation which has long argued for an increased focus on wellbeing and its measurement along with a radical re-orientation of what we value in society. It is interesting to see how, through a succession of changes in central government agendas, the concept of quality of life has been usurped in policy discourses by a focus on wellbeing which is often synonymised with happiness. Like all grand but slippery terms that have high rhetorical value (freedom, justice and democracy, to name a few) the concepts of quality of life and wellbeing have been mobilised in different ways, by different groups, to support different agendas over time. This makes it important to explore what and whose values are represented, which accounts dominate, what is their impact and on whom. Wellbeing indices reflect discourses that hold implicit and explicit messages about the nature of individuals and society, the relationship of citizens and the state, the agents of change and their impact, the causes of problems and their solutions. These discourses influence the ways that wellbeing (and development in general) is conceptualised and measured, which accounts hold greater legitimacy and who holds the authoritative view. They also influence discussions about where the responsibility for wellbeing lies, whether with the state, local government, communities, individuals or with 'society' – a problematic notion in itself.

The current politics of wellbeing in the UK and its pre-occupation with subjective wellbeing measurement is implicated in various discourses related to beliefs about a decline in social values and democratic participation, a breakdown of families and communities, increased individualism and consumerism. Localism is hailed as a solution with more emphasis being placed on local governance, with communities and individuals taking responsibility for local quality of life, wellbeing and sustainability. The UK Coalition Government's Localism Act of 2011 is the legislative manifestation of their Big Society idea. This reflects the 'compassionate conservatism' belief that 'we do not simply need a better vision of society; we need a better understanding of the individual, of what it is to be human' (Norman 2010: 8). These ideas are shaped by, and shape in turn, wider views about what wellbeing is or should be. The current interest in subjective wellbeing, and the way it is discussed, are closely meshed with these ideas of democratic renewal, active communities, individual resilience and volunteerism. The concept of wellbeing is now an inherent part of discourses concerning the responsibilities of citizens and the role of local governance in the current economic crisis.

These discourses are not new, although their specific character and context are unique. The Local Government Act of 2000 was a key vehicle for New Labour's reform of local government. It placed an increased emphasis on community governance and sent a clear message that local government was no longer viewed as a discrete entity providing specific services. Rather, it was the 'lead partner' in a local community, wielding the 'power of wellbeing' to deliver a better quality of life (DETR 2000). At the time, discourses of wellbeing and quality of life were entangled with those of sustainable development, which was often presented as 'quality of life for everyone, now and in the future' (Defra 2005) or 'social, economic and environmental wellbeing' (DETR 2000). These existed alongside narratives of public health that in turn tended to conflate ideas of wellbeing and health (Ganesh and McAllum 2010; Fleuret and Atkinson 2007). As wellbeing rose up the political agenda, national government commissioned several studies to explore and refine the concept (for example, Dolan *et al.* 2006b; Marks *et al.* 2006b) and the Whitehall Wellbeing Working Group developed a statement of common understanding that defined wellbeing as a positive mental state enhanced and supported by various social, environmental and psychological factors (Defra 2007). In the government's drive for a clearer definition of wellbeing, internal struggles over its meaning represented 'stakes in the ground' for particular departments or agendas (Ereaut and Whiting 2008: 13). But despite various discursive framings of wellbeing and the tussles in arriving at a definition, policies aiming to improve wellbeing often focused on local communities. For example, The Local Wellbeing Project, which developed the first set of local wellbeing indicators, gave great weight to 'place-shaping' at neighbourhood level (Steuer and Marks 2008; Mguni and Bacon 2010). Recent work, however, shows that wellbeing is also variously framed and constituted at the level of local governance by local authority partnerships (Atkinson and Joyce 2011).

This book concentrates on the role of local governance in developing definitions and measurements of wellbeing in the context of shifting relationships between citizens and the state. In Part I, I set out the wider philosophical and historical context of wellbeing and its measurement. Since the 1960s significant advances in conceptualising and measuring wellbeing have been made yet, depressingly, many commentators feel that despite this accumulation of knowledge we still cannot address fundamental issues such as global inequalities and impending environmental crises. Critiques of GDP and the pursuit of alternative measurements of progress are not new and we need to learn from previous experience (Michalos 2011). With the aim of promoting wider debate, I offer an overview of the field. I also set wellbeing measurement in the context of other issues implicated in local policymaking: the relationship between wellbeing and sustainability, issues of power and participation, and the role of knowledge and evidence. While some of these issues are universal, my focus here is on their application by western liberal democracies, particularly the UK and specifically local governance.

Part II describes a case study of the attempt by one local authority in the northeast of England to define and measure local wellbeing. It is a context-based exploration of the issues set out in Part I which examines everyday contemporary local policymaking processes. The account is semi-autobiographical; I was a participant observer within a local authority from 2004 to 2007, when the concept of subjective wellbeing was becoming prominent in national policy discourses, and during a process of widespread local government restructuring. I was involved in the practical work of defining and measuring wellbeing, engaging in debates and aiming to influence policy, while simultaneously reflecting on how I and others were and are implicated in various conflicts including how to frame quality of life, wellbeing and sustainable development. I present a local wellbeing framework, which was developed through qualitative research, and assess its significance in comparison with other accounts of wellbeing. Looking particularly at the institutional discourses and power relations involved, I also describe how the framework was created, and consider how these may determine its influence. The final chapter of the book pulls together the key messages from the literature and empirical case study and reviews them in the light of the contemporary political context.

The truth about wellbeing and the trouble with measurement

This book looks at the political project of wellbeing measurement through the lens of discourse theory. This is a wide field and I have not attempted to review here the different ways in which discourse is conceptualised and used as a method of social and political analysis. Essentially, a discourse is a shared understanding of the world, produced by and communicated through language (Dryzek 1997). A tree is a tree; it exists, it has a physical reality with a particular mass and form. It is also socially constructed, in that different groups of people may attribute different meanings to it. It may be an awesome part of nature, for others it may be wood and livelihood, for others it may be an impediment to the ploughing of a field. It could be several of these or something else but it can never be viewed outside meaning, there is no 'view from nowhere'. In talking of the tree, each of these groups will use particular language to achieve their desired ends (the protection, ownership or removal of the tree, for example). Discourse theory does not deny the existence of factual reality, rather it stresses that this reality always has meanings and that these meanings are often so powerfully constructed that they can, and often should, be treated as factual reality (Howarth 2000; Fairclough 2003). The philosopher Michel Foucault offered an explanation of how discourses work to construct social realities by creating rules about what is considered to be the 'truth' (Foucault 1991b). He believed that certain sorts of knowledge, particularly scientific knowledge, are accorded a privileged place in societal views of what constitutes the truth. The recent rise of political interest in wellbeing signals a new set of discourses. Although wellbeing is an old term, a new wave of talking about and understanding wellbeing is having a clearly discernible influence on central and local government, the voluntary sector and civil society. As I will show in the case study, contestation over meaning is an important factor in discussions about wellbeing and its measurement, and this process necessarily involves a struggle over what constitutes legitimate knowledge.

The wellbeing measurement research currently being developed in the UK and elsewhere concentrates on devising statistical indices for different wellbeing dimensions. This is important for all sorts of reasons already rehearsed in the literature, some of which I review in Chapter 2. These indices are required to show wellbeing patterns across different areas, different groups of people and between different dimensions and stages of life, and may be applied at a local, national or international level. They need to be designed with robust comparison as a main criterion which together with resource and technical limitations, places particular constraints on what dimensions of wellbeing can be measured and how. Although increasingly sophisticated concepts of wellbeing are producing more sensitive indicators (for example see Huppert *et al.* 2009) nevertheless such statistics can only ever be abstracted versions of life. If we want to measure something complex we normally have to split it up into dimensions, categories or domains and we need to measure many of these to get a fuller picture. For example, how do we measure the quality of a painting? We might assess its economic value, we might measure its physical size, we might categorise the era it was painted in, the medium used, the type of art, the nationality of the artist, the subject of the painting. We might go on to consider the range of pigments used, the compositional factors, the symbolic components. However, by looking *only* at this information, rather than at the actual thing itself, we could not possibly understand how these dimensions relate to produce this painting. How could we tell if this painting was mediocre or a work of art? This evaluation is based on direct experience, aesthetic judgement and emotional response set within a social debate that creates collective meanings about art which are underpinned and reinforced through cultural context and everyday practices. Of course judgements about what is or is not a work of art are much contested and not necessarily democratic; elitism and power are taken-for-granted factors. I believe the same goes for wellbeing. That is why many wellbeing indicator researchers are driven to seek increasingly sophisticated ways of thinking about and measuring wellbeing:

> Besides looking for models in machinery with discrete parts, wellbeing researchers should be looking at the arts of cooking, weaving, painting, music and literature where holistic thinking and orchestrated designs are known to produce qualities (including information) not present in their distinct parts.

> (Michalos *et al.* 2011: 12)

Their efforts are often motivated by a strong social justice ethic. Nowhere in this book do I seek to devalue their vital work. In a policy world dominated by discourses of 'hard' evidence, policy actors have to back up their decisions with statistical evidence that represents the interests of all concerned as fully as possible. Given the difficulty of such a task, critiques risk sounding churlish. I am not critical of the science of wellbeing measurement itself. My concern is that this type of measurement may be pursued *at the expense of* other forms of knowledge and judgement. I believe that despite commitments to public consultation, scientific expertise is being promoted as the main basis for defining wellbeing, and this may be shutting down rather than opening up debate. A scientifically derived concept is emerging as the ultimate 'true' and 'real' account of wellbeing, rather than simply as one set of evidence to inform debate. This is illustrated by a 2011 academic paper entitled 'Doing the right thing: Measuring wellbeing for public policy' by a team of psychologists promoting 'wellbeing theory'. The paper initially sets out wellbeing as a 'multi-faceted phenomenon' but concludes that getting the correct approach to wellbeing measurement will 'help educate both the public and policymakers about *what wellbeing really is*, and about the multiple ways in which it can be cultivated' (Forgeard *et al.* 2011: 99 [my emphasis]). I believe we should be wary of such truth claims about something so 'inherently ambiguous' (Gasper 2009: 6).

While social science statisticians have had enormous influence in raising awareness and debate about poverty and inequality (for example the work of York-based philanthropist Seebohm Rowntree in the early twentieth century) over-reliance on indices can have negative side effects on society. The school league tables in the UK, for example, started from intentions to create equal standards of education across the country by applying a standard measure of attainment. However, this index has been used by parents with means and aspirations to move their children to 'better schools', thereby perpetuating inequalities. Also, what makes a 'good' school depends on how one sees education and its relationship to wellbeing. Some argue that the choice of academic performance as a proxy for educational quality perpetuates an elitism based on intellectual ability and makes children with other skills and qualities feel devalued in society. This may also harm academically able children who find themselves trapped in a culture where academic performance is the only means they have to gain esteem. Furthermore, many school teachers feel they are working in an increasingly target-driven environment that is having a negative impact on their ability and morale. This 'tick-box culture' creates a gap, not only between the rhetorical values of politicians and the reality of their policies, but more insidiously, between what teachers are being asked to say they can do and what they actually believe they can do. The increasing demands for excellence and efficiency, the way this is measured and the competition it creates can make us feel like failures and frauds. Despite the refinement of educational quality measures across two decades, statistical experts conclude that these indicators are 'not fit for purpose', that they have led to the creation of 'perverse incentives' and that we should move away from measuring and publishing them (Goldstein and Leckie 2008). So we should consider carefully which aspects of wellbeing we measure, what proxies we create, what narratives and discourses inform such indicators and what the overall effect of this may be on society.

Practical wisdom in an uncertain world

Wellbeing measurement at the local governance scale is set within a matrix of uncertainty. Global political and social changes combined with technological progress have caused local governments and communities to question their place, identity, responsibilities and expectations (Mulgan 1997; O'Riordan 2001). Beck's classic theory of an increasing 'risk society' is characterised by a radical uncertainty as modernism creates more and more wicked problems (Beck 1992). Uncertainty is the 'inevitable by-product' of greater interventions in nature and 'political conflicts often cannot be resolved simply by providing more knowledge' (Hajer 2003: 188). The impacts of globalisation and the role of the media have eroded politicians' control of policy definitions and changed the role of classical political institutions as the place of decision making, creating a phenomenon that has been termed the 'institutional gap' (Hajer 2003). New governance theory in the UK characterises these shifts in terms of a loosening of traditional governing hierarchies to include complex and diverse networks involving new actors (Rhodes 1997; Stoker 2000; Kjaer 2004). An emphasis on partnership working at the local level is symptomatic of increasing complexity where policies are decided through consultation with a range of stakeholders (Darlow *et al.* 2007). Private sector, third sector and communities are being expected to fill the gap left by the shrinking state and overstretched local authorities. This move from 'government' to 'governance' is seen by many as a threat to transparency, the danger being that certain interests will be by-passed by network decision-making outside the public sector (Stoker 2000). To head off this danger, community participation has been called for to 're-introduce direct accountability' and there has been an increased international emphasis on participatory processes (Kjaer 2004: 15; Taylor 2007). Participatory democracy has moved from the margins of government and been adopted as a mainstream idea. It is also seen as one way to address the fall in voting turnout, a subject of concern in many established democracies which many perceive as a symptom of democratic crisis (Zittel and Fuchs 2007; Miller *et al.* 2000). In this profound uncertainty we need our democratic institutions more than ever and it is important that we work hard to retain and improve a democratic decision-making culture.

Participatory democracy is central to notions of wellbeing. The increasing focus on ever more sophisticated measurement tools for wellbeing does not necessarily help policymakers and may also abstract wellbeing from context-based, common-sense understandings. The fact that it means different things to different people is often seen as problematic in defining and measuring wellbeing, but it is often neglected that, crucially for democracy, it also means something to everyone. As wellbeing is increasingly framed as a scientific rather than a political project in policy discourses, common-sense understandings of wellbeing may be undermined and citizens may become less confident about which debates they are qualified to enter. In this book I consider the concept of measurement in some of its wider, older senses: 'to take the measure of'; 'to measure up against'; 'to give full measure to'. These phrases suggest not statistical accounting but *practical judgement*. They suggest a rounded view, a common-sense assessment gained not only from science and statistics but also from experience of ourselves and others in the world, a knowledge that is embedded in context and culture. Metaphors for practical judgements (e.g. weighing up, taking stock of, accounting for) often evoke the world of measurement implying a direct link between quantification and decision making. In the modern world where a profusion of accounts and statistics reflects a complex of interconnected global issues, this link is more obscure. Faced with such complexity and uncertainty, we feel ill equipped to make decisions and therefore, perversely, require ever more evidence and statistics, which become counter-productive in that they complicate the decision-making process. And still we are accumulating more data, constructing indicators and policy frameworks. Yes, evidence-gathering is a necessary and dynamic response to a changing world but the search can distract from the fact that defining and measuring wellbeing is at root a political project.

The new discourses of wellbeing make much of Aristotle's notions of 'oikonomia' and 'eudaimonia' in their discussion of economics and its relationship to human flourishing. They have less to say about his ideas on practical wisdom or 'phronesis' which tell us how we should make political judgements regarding the good life. Phronesis is the accumulation and application of extensive experience in the real world of many different situations. Political expertise therefore, is not in the ability to apply rule-based objective knowledge, but in the ability to apply one's experience to different contexts. Phronesis differs from analytical knowledge ('episteme') or technical knowledge ('techne') and it is interesting that whereas the last two have found their way into modern research language (as epistemology and technology) phronesis has no modern derivative and yet Aristotle considered it to be the most important intellectual virtue for political and social inquiry (Flyvbjerg 2001: 3). Flyvbjerg notably transfers the concept of phronesis to social science research methodology by conducting a detailed context-dependent analysis of local governance to demonstrate the relationship of rationality and power. His classic study of the Danish municipality of Aalborg over fifteen years showed that the idea that policymakers use evidence-based rationality to drive policy is a fallacy (Flyvbjerg 1998). I will argue throughout this book that wellbeing measurement should recognise the realities of political culture and policymaking, rather than assuming it is a technocratic rational process. Indicators can be produced by, but cannot substitute for, democratic engagement, intelligence, understanding and experience on the part of community leaders and policymakers.

Key issues and real lives

In trying to develop wellbeing measurement for a local area, four linked but separate areas of inquiry should be made. The first is how to define wellbeing or quality of life; the second is how wellbeing and sustainability relate to each other and the trade-offs to be made; the third is who to involve in these discussions; the fourth is how indicators influence policy to improve wellbeing. In order to illustrate these issues I set out a summary of discussions with two residents during the case study:

Case Study 1.1

John is in his forties, he has a physical disability and uses an electric mobility scooter to get around. He lives in a council-owned house in an area described by the council as 'disadvantaged'. He is not in paid employment and claims benefits. He lives on his own and says he could not do without his TV. Although his health limits what he can do, he thinks it is important to go out every day and be active. He goes to church and feels that his faith is very important to him. He believes strongly in helping others and does bits of shopping for his neighbours and they help him out too. It's important to him to feel 'safe and peaceful', to be able 'to have a rest in peace and quiet'. He says he generally enjoys life and the things that make him particularly happy are gardening, growing flowers and his pets.

Ann's husband died and she is bringing up their daughter Carly, 12, on her own. Ann has not found a way to deal with her husband's sudden death two years previously, 'he was the love of my life,' and she 'struggles with depression' and tends to stay in the house at night and drink 'to numb the pain'. Ann feels that Carly isn't having the best start in life and would love to be able to take her out more but she can't afford this. Ann works full time in a factory job and resents the amount of tax she has to pay. She thinks this is unfair when she earns so little compared with others. She would be scared to leave her current employment for a better paid job because she's been there a long time and has built a good relationship with the other women at work, 'we have a good laugh' and that is her only social life. She is too tired in the evening to do anything although she says that 'staying in gets me down, depressed'. She feels she is 'a negative person anyway due to my childhood.'

John, by most objective indicators commonly used, would be viewed as having a relatively poor quality of life. He has no paid work, has long-term health problems, lives alone in a part of town that according to objective statistics is 'deprived'. Yet he does much unpaid work that is important to his local community, he finds quiet satisfaction in relatively simple things and has good relationships with his neighbours and fellow churchgoers. He self-assesses as being quite happy and having a good quality of life. Ann and Carly, on the other hand, don't have any easily visible or measurable problems. Ann is in full-time employment, Carly is doing OK at school, they live in a quiet neighbourhood and from the outside, they are an ordinary family getting on with their lives. How can we calibrate the quality of life of this family who are living a life like many others but which they feel is not the best life they could be living? How could we measure their lack of opportunities/capabilities and their unhappiness? The point of these examples is not to say that measurement has no value; on the contrary, I argue later that it is important. The examples are intended to show why we need to ask the following sets of questions about the measurement of wellbeing for public policy.

Philosophies of wellbeing

First, it is well established that wellbeing[6] is multi-dimensional and complex, and sensitive to both cultural and individual interpretations. Economic systems, societal culture, relationships, environment, health, personality, upbringing, aspirations are some of the aspects of human life mentioned here, and there are many others, which interact in a unique way for each person. Therefore it is important to engage with people's own account of their wellbeing. However, people have adaptive preferences (Elster 1983) and they may become accustomed to a lower quality of life (assessed against 'objective' indicators), so should we trust people's own account of their wellbeing? But if not, whose account should we trust? Perhaps one based on an 'objective' account of wellbeing? But then how would we vary this for different people's needs and different cultural and societal conditions? Moreover, where should 'objective' values come from? Considering this diversity and complexity, is it possible for a coherent and manageable philosophy of wellbeing to emerge that can inform indicator development for public policy? How can we move a collection of individual notions of wellbeing and theories about how to increase it, into the strategic policy development arena? What are the differences between an individual and a societal conception of wellbeing? What would be the implications for measurement?

Relationship of wellbeing and sustainable development

Second, understandings of wellbeing co-exist and are sometimes synonymised with the concept of sustainable development, which has emerged as the 'overarching framework' to which public policy should refer (Rydin *et al.* 2003). But how do concepts like intergenerational justice and ecological protection merge with the ideas of John and Ann about the quality of their lives? Quality of life indicator sets are often developed within a sustainable development (SD) paradigm, with a key aim of raising awareness of SD through their development. This conflation of terms represents a fundamental dilemma in the study of locally-developed indicators. Empirical research in the USA and the UK has shown that members of the public are often unfamiliar with the abstract concept of SD (Lawrence 1998; Burningham and Thrush 2001). Therefore, researchers and practitioners trying to engage local people in developing sustainability indicators argue that discussions should be framed around local concerns, using terms and concepts that people can understand, like wellbeing and quality of life (Lawrence 1998; Lingayah and Sommer 2001). Local participation is central to the notion of SD, yet when this participation occurs around local quality of life issues, the outcomes do not necessarily address the key SD questions; the global and intergenerational equity aspects of sustainable development are rarely fully understood or explored. Therefore local indicators have tended to reflect 'immediate and local concerns' with an emphasis on socio-economic issues, underplaying global and ecological issues (Rydin 2007). How can we reconcile local conceptions of wellbeing with the more abstract notion of sustainable development?

A key role for democracy?

The third set of questions to consider is who should be included in discussions to define and measure wellbeing, and how? The actors involved and their social context are crucial factors, because notions of wellbeing arise from political and philosophical values which drive forward ideas about development. These notions may be in part driven by central priorities or by other discourses which promote particular values in society, so which values should we regard as the best? How can local democracy be made to respond to this challenge? The contested nature of these theories demands that local people participate in creating meaning and measurement. As Dryzek (1997: 84) puts it, 'Leave it to the people' but like wellbeing, quality of life and SD, participation is itself a 'problematic and a political concept which…needs extensive debate to ensure that it is properly and effectively understood and deployed' (Buckingham-Hatfield and Evans 1996a: 13). The introduction of the term 'community' complicates debates by presenting yet another contestable notion (Dinham 2006: 182). So the idea of promoting community participation to debate the concepts of wellbeing and SD only seems to add further contestation and complexity. This complexity has produced rich seams of literature on the theory and practice of participation, from 'how to' toolkits for working with local people to develop indicators, to advanced critiques of participation showing a deep cynicism for the practice (for example Cooke and Kothari's (2001) *Participation: The New Tyranny?*). Accordingly, this book concentrates on the power relations involved in creating such indicators through 'community' participation.

What is the role of knowledge and indicators in policymaking?

This leads us to a fourth set of questions: even if some consensus could exist at a local level, how would these notions be turned into indicators? How could indicators be turned into actions that really make a difference to Ann, Carly and John? And if indicators do not help decision-makers to make a difference to individual lives such as these, what is their purpose? The idea that data, measurements and information are needed to inform rational, evidence-based policy is ingrained in the UK public sector, and yet practitioners and researchers are disappointed with the weak link between indicators and policymaking leading to any concrete change (Boulanger 2007; Rydin 2007; Rydin *et al.*, 2003; Levett, 1998; Cobb 2000; Innes 1990). Furthermore it has been argued by philosophers and shown through empirical research that policy and politics works through complex relationships of power, knowledge and rationality (Flyvbjerg 1998; Foucault 1991a). So how are we to take this into account in the practical project of wellbeing measurement?

Summary

This book is organised around these four related areas of inquiry. I explore the theoretical concepts and local understandings of wellbeing; I explore the relationship between concepts of human wellbeing and sustainable development and the tensions between them; I explore the complex issues around creating the forms of participatory governance that can support these deliberations; and I explore the nature of indicators and their role in policymaking processes to promote wellbeing. The four remaining chapters that make up Part I address each of these questions in turn and provide an overview of the complex field of enquiry we need to undertake in order to devise local policy for wellbeing. Part II of the book explores these questions through a detailed case study.

This book engages with both the philosophical and practical challenges of defining, measuring and promoting wellbeing and sustainability. Wellbeing definitions and their measurements both shape and are shaped by their cultural context, social norms and institutions, and this affects the way measurements are developed and chosen and how they are used (or not) (Galloway 2005; Astleithner and Hamedinger 2003; Astleithner *et al.* 2004). This book particularly focuses on the relationship between central discourses and local governance, it looks at the development of indicators and the processes by which they are created, at the local level. This is not a 'how to' book although I believe the case study has practical implications for policymaking. The book focuses on the policy processes, institutional norms and power struggles that the definition and measurement of wellbeing entail. Although the focus on wellbeing for public policy goes hand in hand with attempts to 'measure what matters', I show that not only is 'what matters' up for debate, but also how we measure it. It is how we acknowledge and value that debate and the different types of knowledge and experience which can inform it, that is the real challenge of measuring wellbeing.

2　Human wellbeing and quality of life

I wish you health and happiness
I wish you golden store
I wish you heaven when you die
What can I wish thee more?

This poem was once a popular sentiment sewn into embroidery samplers and quilts. Some may consider it trite. For my grandfather, it must have had a deeper meaning. He wrote it on the back of a postcard and sent it to my grandmother soon after they were married. It was 1945 and he was a prisoner of war in Burma. I imagine that every word sent home counted a great deal. When it comes down to it, most of us can express our ideas of wellbeing quite well, according to what we value for ourselves and our loved ones, in the context of our social situation and culture. Since my grandfather wrote that postcard, the world has changed profoundly in ways he could never have foreseen. In some ways our values and expectations have shifted as well; in others they have endured. As the philosopher Hannah Arendt put it, 'we are all the same, that is, human, in such a way that nobody is ever the same as anyone else who ever lived, lives, or will live' (1958: 7–8).

How can we capture sameness and difference at once? This is one of the fundamental challenges to the quest for a common definition of wellbeing for public policy. How can we arrive at a common definition that can support strategic decisions without restricting our freedom as individuals to choose the way we want to lead our lives? What would this definition look like? Philosophers have wrestled with these questions since Aristotle if not before.[1] We now have an extensive literature on quality of life, wellbeing, life-satisfaction, welfare, utility, happiness, prosperity, human flourishing, human development and so on. This chapter reviews a wide and complex field. Limitations of space and my focus on conceptions of wellbeing for general political purposes necessarily result in some significant omissions. For instance, I do not cover here more specific health-related quality of life measures. For fuller and more learned reviews of the range of quality of life constructs, research and measurement see Phillips (2006) and Sirgy *et al.* (2006). The discussion here focuses mainly on the terms 'wellbeing' and 'quality of life' as these are the ones most frequently used in UK policy literature and discussions regarding measurement. They are often used interchangeably, being intuitively understood generic terms for a wide range of ideas about what constitutes a 'well-lived life' (Dasgupta 2004: 13). This inherent accessibility is part of the common-sense value of these terms in democratic debate. Nevertheless there is a range of possible meanings. I will explore some specific constructs of wellbeing and quality of life in this chapter by considering the roles of philosophy and science over the last fifty years, including within current discourses on 'happiness' measurement and policy in the UK.

What is it to be 'truly' human?

Theories of wellbeing or quality of life cannot be disengaged from theories of what it is to be human and what life is for. Even the most liberal accounts of wellbeing, some of which are reviewed in this chapter, are predicated on assumptions that as humans we are essentially tolerant and reasonable beings, that we can live and let live. Wellbeing theories that claim to be grounded in a universal ethic, will still privilege certain ideas of humanity over others. Of any account of wellbeing, we should ask, what assumptions does this make about human nature and what does it prescribe? Our views on what makes a good life are also informed by a wide range of beliefs about the meaning and purpose of life. The nature of human beings and the meaning of life are of course entwined. If one believes in some form of after-life, the pursuit of wellbeing will also take that into account; recognition of a higher spiritual authority may take precedence over personal autonomy. Wellbeing in this life may be evaluated in relation to the prospect of greater wellbeing in the next. From a more existentialist viewpoint, individual autonomy may be the primary criterion for self-realisation or wellbeing. If one believes that human functioning is inextricably connected to society, the health of social groupings, or the common good may take priority over individual preference. If one believes that human life is part of a fragile ecosystem on which wellbeing depends, then the good life may be determined by the health of that whole system. This is not to say that we cannot believe some of these things (or others) simultaneously or that our behaviour always follows our beliefs, but those are complex discussions, outside the scope of this book. My point is that all theories of wellbeing and its measurement are underpinned by values and beliefs, whether explicitly stated or not, about the nature of humanity and the meaning of life. So while it is common and understandable for researchers to claim 'neutrality' in measuring wellbeing, the idea is really a nonsense. Wellbeing studies should therefore consider the context within which discussions about measurement and policy occur, and which beliefs and values they privilege, exclude or undermine. One means of doing this is through discourse analysis, which examines language, social practices and power relations in order to evaluate what is being promoted in a particular account of wellbeing and, crucially, what impact it has, and on whom.

Critiques of utilitarianism and GNP

During the early part of the twentieth century, the dominant idea of human welfare was defined by material wellbeing. Subsequently, economic growth, measured by gross national product (GNP) and now predominantly by gross domestic product (GDP)[2] became a proxy for social progress (Offer 2000; Veenhoven 1996). Although economists (including Simon Kuznets, the creator of the GNP prototype) have often pointed out that it was never intended to be a measure of welfare, politicians have used it in this way, placing great importance on increased economic growth as a welfare goal in itself (Dasgupta 2004; Levett 1998). This link between economic growth and wellbeing reflects utilitarian ideologies of maximising welfare[3] in society. Utilitarianism has its roots in the philosophy of social reformer Jeremy Bentham (1748–1832) who sought to promote policies that improved utility for the greatest number of people in a society. According to Bentham, utility or happiness meant the presence of pleasure and absence of pain; as people themselves are the best judges of what gives them pleasure and pain then they should be free to satisfy their own preferences.[4] One of the best ways to do this is to provide people with the means to exercise these preferences, income being a key factor, hence the subsequent meshing of utilitarianism with GDP approaches to welfare. The idea of individual preference has dominated wellbeing studies in economic theory for over 100 years (Dolan *et al.* 2006b). The measurement of GNP/GDP became inextricably linked with these notions of welfare and became a powerful, internationally institutionalised indicator which incorporated the notion of human development. The prioritisation of economic growth fuelled by the aspiration to raise GDP has long underpinned Western public policy.

To make as many people as possible happier seems a reasonable goal but for critics of utilitarianism (and there are many), the aggregation of welfare levels in society can hide glaring inequalities and the interests of minorities can be consistently overlooked (Dasgupta 2004; Nussbaum 2000; Sen 1980). In addition, if increased happiness is the goal, according to utilitarian principles unequal pay between men and women, for example, does not matter so long as women are content with the situation (Robeyns 2005: 96). Therefore utilitarianism comes under strong critique from those who are concerned with social justice because 'it cannot directly rule out slavery, or the oppression of women, or the misery of the poor' (Nussbaum 2005: 35).

In addition, a reliance on preference satisfaction as a mechanism for wellbeing lacks explanatory clarity for the psychological, cultural and social complexity behind choices; it does not distinguish between a wide range of concepts which might affect people's decisions, such as 'belief, desire, perception, appetite, emotion, impulse, inclination, intention' (Nussbaum 2005: 34). Studies in social economics, for example, have shown the importance of norms and institutions associated with family, place and tradition in how people make decisions regarding their livelihoods (Oughton *et al.* 2003). Furthermore, preference satisfaction is inefficient as a mechanism for wellbeing as we cannot always satisfy our preferences in the same way that we can satisfy a need (O'Neill 2011). Preferences can be insatiable, especially when coupled with an increasing choice of consumer goods. Instead of increasing happiness in society Easterlin (1974) showed that the post-war rise in general affluence had not delivered a similar rise in levels of life satisfaction (the so-called 'Easterlin paradox'), and increasingly, utilitarian theories associated with preference satisfaction and economic growth were found to be lacking.

Dissatisfaction with GNP as an indicator for human development grew alongside an anti-materialist movement in the 1960s (Offer 2000). New theories about the ideological basis and measurement of wellbeing started to emerge:

> It was in was this period of prosperity, when for the first time doubts were raised in the highly developed western societies about economic growth as the major goal of societal progress. The 'social costs' of economic growth and 'public poverty' as the other side of the coin of 'private affluence' got public attention and received prominence in political discussions. There was increasing doubt whether more should ever equal better, and it became a public claim to prefer quality to quantity. The concept of 'quality of life' was born as an alternative to the more and more questionable concept of the affluent society and became the new, but also much more complex and multidimensional goal of societal development.
>
> (Noll, 2000)

Lyndon B. Johnson, US president at the time, picked up on this mood in his 'Great Society' speeches during 1964 which cautioned against unbridled growth for its own sake and called on citizen initiative and traditional values to steer a new vision of society. Johnson is credited with coining the phrase 'quality of life' in its modern form (Rapley 2003; Noll 2000). Robert Kennedy put the point eloquently, later that decade:

> Too much and too long, we seem to have surrendered community excellence and community values in the mere accumulation of material things. Our gross national product – if we should judge America by that – counts air pollution and cigarette advertising, and ambulances to clear our highways of carnage. It counts special locks for our doors and the jails for those who break them. It counts the destruction of our redwoods and the loss of our natural wonder in chaotic sprawl. It counts napalm and the cost of a nuclear warhead, and armored cars for police who fight riots in our streets. It counts Whitman's rifle and Speck's knife, and the television programs which glorify violence in order to sell toys to our children. Yet the gross national product does not allow for the health of our children, the quality of their education, or the joy of their play. It does not include the beauty of our poetry or the strength of our marriages; the intelligence of our public debate or the integrity of our public officials. It measures neither our wit nor our courage; neither our wisdom nor our learning; neither our compassion nor our devotion to our country; it measures everything, in short, except that which makes life worthwhile. And it tells us everything about America except why we are proud that we are Americans.
>
> (Robert Kennedy Address, University of Kansas March 18, 1968)

Although this has strong echoes with the 'Big Society' and wellbeing discourses of UK prime minister David Cameron (who has quoted Robert Kennedy), the Great Society discourses of quality of life were set within a context of post-war prosperity, not economic crisis and public service cuts. The focus was on welfare provision and social reform to distribute the benefits of economic growth more widely. In the civil movements of the 1960s disenfranchised groups found their voice and claimed their fair share of economic opportunity and social mobility. In the UK, Harold Wilson's Labour government prioritised major spending on welfare provision, social housing, education and health and legislated busily to improve civil rights and social freedoms.

The social indicator movement

These movements went hand in hand with various attempts to better account for quality of life in national indicators, resulting in 'extended national accounts' and the 'social indicator movement' (Offer 2000). Economists worked hard to factor non-market values into the individual preference model and thereby extend the model of national economic accounting (Jordan 2008; Offer 2000). At the same time, social scientists endeavoured to capture aspects of life important to wellbeing which had not previously been measured. The 'social indicator movement' or 'quality of life movement' spread throughout Europe and the USA, characterised by a proliferation of social reporting projects and the gathering of a huge amount of statistical information (Rapley 2003; Offer 2000; Cobb 2000; Gasteyer and Flora 1999; Michalos 1997).[5] Its promoters sought to quantify this new complex idea of 'quality of life' and to reproduce the success that economic indicators had previously enjoyed in terms of directing policy. In Europe, this trend is exemplified by the first Swedish Level of Living Survey in 1965. Responding to the recognition that 'GNP is an insufficient measure of the wellbeing of citizens' this survey commissioned a study of 'the distribution of welfare in non-monetary terms' using 'objective' indicators such as life expectancy, employment and education levels (Erikson 1993: 68).

In the UK, the annual *Social Trends* report was first published in December 1970 and contained over 200 pages of charts, tables and diagrams intended to measure social as well as economic progress. Muriel Nissel, its first editor, gives an engaging account of its launch at No 10 Downing Street to a rendition of chamber music and massive media coverage. That the report was responding to a widespread public call for change is evident in its being voted runner up for the title of the best new reference book that year (apparently being let down only by the quality of its binding) and included on the Christmas reading list in several Sunday supplements. The report for the first time gathered together in one volume a wealth of social statistics, to give a picture of Britain that had not been seen before: it was 'above all concerned with people' (Nissel 1995: 492).

However, authors reviewing this period of social reporting record the disappointing impact on policy achieved by this vast amount of data. They offer three explanations. First, there was a vain expectation that data in itself would inform policy, and the more data the better. This expectation failed as the bewildering array of indicators and information was not linked to any coherent framework or theoretical analysis of what constitutes quality of life, or of what needed to be achieved for whom, and how (Innes 1990; Offer 2000; Cobb 2000). Accordingly, although there was a growing recognition of the limitations of GNP, it remained a single indicator linked to embedded (albeit flawed) theories of how to create wellbeing, and continued to influence and justify policy decisions which prioritised economic growth. Second, there was a growing realism that measurement of social phenomena was more complex than previously imagined, and researchers became disillusioned with inadequate 'surrogate measures' or 'proxies' for public goods such as education and health (Nissel 1995). Third, the political climate changed in the late 1970s and 80s, as economic difficulties and the rise of right-wing neoliberal ideologies in the Thatcher/Reagan era unseated the social indicators movement (Rapley 2003; Nissel 1995). As Offer (2000: 12) says, social indicators relied on a 'social-democratic consensus' but 'by the time social indicators were delivered, the impetus of social democracy was spent'. In the UK, work that had been done in the late 1970s on measuring the wellbeing of minority groups became politically sensitive, and fell by the wayside (Nissel 1995). Prime Minister Thatcher famously said 'there is no such thing as society, there are individual men and women and there are families'[6] and championed the view of active citizens responsible for their own welfare. It was during this era of 'conservative ideology with its celebration of the market and the responsibility of individuals', that interest in quality of life on the scale of the individual arose (Rapley 2003: 8). The so-called 'American model' of quality of life measurement became primarily based on subjective indicators at the level of individual citizens, for example, measures of personal satisfaction or happiness (Noll 2000).

Quality of life and the enterprise culture

The term 'quality of life' became re-embedded in the discourses of government at this time but it was now linked to the rise in the enterprise culture. The concept of quality of life was deployed in all manner of public discourse during the 1980s and 1990s, from advertisements for household goods to healthcare strategies. It became enmeshed with the idea of 'subjective beings' who can buy/choose from different quality of life options in all areas of life (Rose 1992). For instance, Rapley (2003) shows how in ten years spanning the emergence of the Thatcher enterprise culture in the UK, people with learning disabilities had been transformed in government rhetoric from a homogeneous group of 'the mentally handicapped' who needed nursing and care, to individual citizens with the potential to make a contribution to society. The idea of quality of life was thus connected both with new imperatives in care provision and with a new construction of the learning disabled as consumers who exercise choice and responsibility for their own care and support. This is strongly connected to the neoliberal economic project of creating autonomous individuals, independent of the state, responsible for their own quality of life and able to create the means for enriching it. But much deeper than that, it reflects how we view the self:

And this image of an 'enterprising self' is so potent because it is not an idiosyncratic obsession of the right of the political spectrum. On the contrary, it resonates with basic presuppositions concerning the contemporary self that are widely distributed in our present, presuppositions that are embodied in the very language that we use to make persons thinkable, and in our ideal conceptions of what people should be.

(Rose 1992: 141)

These discursively constructed notions of the self are of course continually produced and re-enforced by market mechanisms (consider the ubiquitous advertising mantras 'treat yourself', 'you deserve it' and 'you're worth it'). However, they are also intimately connected to ethical concepts of rights, freedom, justice, equity and democracy which have long historical roots and were mobilised forcefully in the movements of the 1960s. Although many blame neoliberalism for the current social, environmental and economic ills, this is too simplistic. The meshing of market-based mechanisms with deeply held ethical norms has produced a particular idea of the self; unique and deserving with an identity distinct from, but of equal worth to, all others. This dominant construction has become so embedded in our daily lives in the west that it is almost impossible for most of us to conceive of human life and the quality of our lives differently, and indeed we may deem it unethical to do so. This is precisely why such 'realities' should be questioned, and why they are so difficult to question. I return to this in the next chapter when I consider the challenge of trying to integrate the concepts of sustainable development and wellbeing.

An increasingly individualistic construct of quality of life provoked a rise of interest in subjective indicators to measure individual self-reported experience of wellbeing. In 1983, these subjective indicators were included in the *British Social Attitudes Survey*, which highlighted the importance of factors such as marriage, friends, family life and the weather, as well as the already-reported jobs, health and housing concerns (Nissel 1995). These measures provided new dimensions to the notion of wellbeing, focusing on individual experience and preference. Social researchers in Scandinavia however, rejected this model. For instance, Erikson and Uusitalo (1987) argued that it should be resources rather than satisfaction or self-assessed needs that should be measured, for through these resources people acquire their own welfare. In the 'Scandinavian model' thinking focused on objective indicators of level of living or quality of life of society as a whole, such as employment rates, education and life expectancy, and rejected measurements of personal satisfaction (Noll 2000).

Towards a philosophy of social justice

New ideas in philosophy emerged in parallel during the 1970s and 80s which challenged existing utilitarian views of human welfare. A significant contribution came from political philosopher John Rawls, whose influential work, *A Theory of Justice*, was published in 1971. This book argued for 'justice as fairness' as opposed to maximising aggregate happiness in society. He rejected average 'utility' as the goal of policy and instead proposed that each individual should be able to enjoy 'primary goods'. These are certain inviolable basic rights and freedoms, like free speech, which should not be compromised by attempts to maximise happiness (Rawls 1971; Cohen 1993).[7] However, he recognised that in a complex society inequalities exist, so he also argued that society should be organised so that *some* inequalities were to be tolerated if they improved the absolute position of the worst off (Rawls 1971). For example, unequal wealth creation and distribution may be tolerated provided some of the increased wealth improves the lives of the poorest. Rawls suggested using an index of social position determined by income to identify the 'worst off' in society. He believed that the happiness of a society was not directly measurable, and that it should not be the basis for justice, but that his theory of justice, if applied, would increase happiness. However, his focus on income was criticised heavily as too crude a proxy for the distribution of other primary goods, which included 'the social basis of self-respect' (Nussbaum 2005: 38).

Rawls' work influenced a new generation of philosophical thought on human wellbeing developed around theories of equity and justice. Important debates arose about the sort of equity to be promoted: should we aim for equality of welfare (happiness) or for equality of resources? This debate was notably explored by Dworkin (1981a, 1981b) who argued that equality of resources should be prioritised over equality of welfare. Amartya Sen, the Nobel prize-winning economist, promoted the concept of 'capabilities' as the basis of human flourishing (Sen 1980, 1993) arguing that equality of resources and equality of welfare are both flawed in achieving social justice. Sen's ideas have had wide impact particularly in the development field. This approach was influential in the creation of the Human Development Index (HDI) in 1990, now an internationally used composite indicator which includes longevity, literacy and education as well as GDP to generate national comparisons. This has been influential in raising awareness of global inequalities and the plight of the world's poorest.

The capability approach

Capability theory proposes that equality in what people are able to be and to do is a more valuable measure of quality of life than a model based on either equality of resources or welfare. Sen proposes this as a 'natural extension of Rawls' concern with primary goods, shifting attention from goods to what goods do to human beings'[8] (Sen 1980: 218–219). Sen locates the key to wellbeing in the freedom to fulfil human potential, rather than in the technical possession of rights or resources. He distinguishes between capabilities and functionings. Put crudely, functionings are what people do, capabilities are what people are able to do. If we concentrate on functionings as a definition of wellbeing, we may become overly prescriptive. For example, to eat is a basic human function; the opportunity or ability to eat is a capability. It may be important for someone to choose not to eat in certain circumstances, such as a fast or hunger strike, because they are fulfilling another need important for human wellbeing, freedom of religious or political expression. However, no-one should be prevented from having the opportunity to eat and therefore it is opportunity or capability which is the most important focus for wellbeing accounts.

There are many overlaps between functionings, capabilities and needs but it is important to retain a philosophical distinction because the general provision of a public good, a right or a resource is meaningless if a person has not the knowledge, ability, confidence or freedom to access or exercise it. The capability approach is appealing because it takes into account 'the economic, social, political and cultural dimensions of life' (Robeyns 2005). This approach does not reject resources as unimportant, clearly they are not, but holds that measuring these alone gives a distorted picture of human wellbeing. For instance, a person with disability may need more resources than a person without disability to achieve a similar quality of life (Alkire 2005). Likewise, using only individual happiness or satisfaction as a measure of equity can also cause distortions because people have 'adaptive preferences' that mean they can become accustomed to poor conditions and consequently lower their expectations (Elster 1983). For these reasons many capability theorists argue for 'objective' indicators such as levels of education and life expectancy, to be used to highlight the differentials in opportunities.

> The problem with an approach based on people's own assessment of their degree of satisfaction is that it is partly determined by their level of aspiration, that is, what they consider their rightful due. This means that measuring how satisfied people are is to a large extent equivalent to measuring how well they have adapted to their present conditions.
>
> (Erikson 1993: 77)

Researchers and policy actors are becoming increasingly interested in the capabilities approach. This, however, is a complex concept with different cultural meanings. This poses considerable theoretical and practical problems. The challenges of specifying particular capabilities for a good life highlight a critical struggle:

> The search for a universally accepted applicable account of the quality of human life has, on its side, the promise of a greater power to stand up for the lives of those whom tradition has oppressed or marginalized. But it faces the epistemological difficulty of grounding such an account in an adequate way, saying where the norms come from and how they can be known to be best.
>
> (Nussbaum and Sen 1993: 4)

Due to this risk of imposing inappropriate or unacceptable norms, discussions of local democracy and participation are central to accounts of quality of life in the capabilities literature. For this reason, Sen has always veered away from a prescriptive account of what quality of life should look like, preferring to leave this to democratic decisions in each area. However, Martha Nussbaum has taken a different view and has produced a much discussed list of ten central human capabilities: life; bodily health; bodily integrity; senses, imagination and thought; emotions; practical reason; affiliation; other species; play; control over one's environment (see Appendix B for a full account). Nussbaum argues for the need to set out a substantive account of fundamental capabilities that are 'a minimum account of social justice' for political purposes. She recommends a 'political focus on the individual' because political accounts should aim to support each person to function in a 'truly human way' with human dignity (Nussbaum 2000: 72). It provides the 'underpinnings of basic political principles that can be embodied in constitutional guarantees' (Nussbaum 2000: 72–76). Her list determines a basic minimum that people of otherwise different values can agree to, an 'overlapping consensus' (using Rawls' term) or common ground upon which all can stand. Because her list is based on capabilities and not functionings (what people are able to do rather than what they actually do) it avoids over-prescription and allows people to subscribe to options that they recognise as beneficial to others, even if they do not wish them for themselves.

This approach is influenced by Aristotelian concepts of 'eudaimonia' or flourishing, and is an example of a 'perfectionist' or 'classical' account which aims at a broad conception of what the good life is or should be. Such accounts are often criticised for being paternalistic, unlike 'liberal neutrality' accounts which propose a basic minimum of public goods to allow people to chose their own good life in a pluralist society. However, Aristotle himself argued that a basic conception of the good life is necessary to create conditions whereby plurality can flourish (O'Neill 1993). There is also an important argument to be had about the point where the argument for providing minimum public goods becomes a perfectionist account of the basic social minimum for a good life. In a defence against criticisms of imposing a paternalistic set of values, Nussbaum describes her list as general enough to accommodate cultural differences, and argues that the detail will still need to be decided through local democratic process. She argues for universal values by pointing out that cultural ideas have now spread across the globe and it may be hard to define what exactly 'local' means (Nussbaum 2000: 49). Furthermore, local ideas may be values of a powerful elite imposed on others, or of a majority which excludes minority views; many existing local value systems are already paternalistic and limit what certain groups can and cannot do. 'Why,' she asks 'should we follow the local ideas, rather than the best ideas we can find?' (Nussbaum 2003: 40). I pay particular attention to Nussbaum's approach because I compare her philosophical and universal account with a wellbeing framework built on 'local' views described in the case study later in this book (see Chapter 8 for comparison and analysis).

The capability approach has a 'special character' because it focuses on 'potential rather than actual outcomes' which makes measurement challenging (Gasper 2007b: 478). How can we measure flourishing or potential? There are considerable difficulties in applying a capability approach. Nussbaum made a start by setting out a list of central capabilities, and Anand *et al.* (2005; 2007a; 2007b; 2009) have tested and refined a list of indicators that is based on this list and applicable to British culture. This included survey questions such as 'Do you have as much love and support as you need?', 'Do you have the opportunity to go on holiday?' Some local authorities have begun to put this type of question into their annual surveys.[9] However, it is unclear what could be done with this information in terms of public policy. Moreover, in spite of the theoretical distinctions between a focus on happiness, resources or capabilities, in practice these things are not so easily distinguishable.

The needs approach

Closely linked to, but distinct from, the capabilities approach, is the human needs approach. Maslow's (1943; 1954) theory of needs is perhaps the most well known example of this. Maslow ordered them into a series of layers, with physiological needs at the bottom which must be met before working upwards though safety, love/belonging, esteem to the pinnacle of self-actualisation. He refined this basic model to include other layers but essentially his theory is that we tend to meet our needs in a particular order. This hierarchical approach has been very influential but the prioritisation of physiological needs risks being conflated with a concept of 'basic needs' (as I show in the case study later) which may allow policymakers to have a very 'thin' view of wellbeing, particularly for the poorest. Other needs, like that for love and support, are viewed as important but in a sequential fashion, after basic needs are met. This is in contrast to accounts like Nussbaum's, which puts forward a set of capabilities where the presence of each is necessary to lift an individual above a minimum socially-accepted threshold for quality of life.

Doyal and Gough's (1991) *A Theory of Human Need* is perhaps the best theorised and discussed contemporary account of needs. Needs are described by Doyal and Gough (1991: 55) as 'a particular category of goals which are believed to be universalisable' and if not met 'serious harm of some *objective* kind will result' [my emphasis]. As such, they forcefully argue that if a person is in need it is the duty of society to meet that need. Like Nussbaum they consider it important 'to clarify and defend those universal human interests which alone can underpin an emancipatory and effective political programme for all men and women' (Gough 2003: 3). Doyal and Gough present a dual structure of needs: primary needs, which must be met at an optimum level; and intermediate needs which must be met at the minimum level to produce the optimum level of primary needs. Primary needs are physical health and autonomy of agency. Intermediate needs are: nutritional food and clean water; protective housing; non-hazardous work environment; non-hazardous physical environment; safe birth control and child bearing; appropriate health care; secure childhood; significant primary relationships; physical security; economic security; appropriate education.

Their account is critiqued for being a rather 'stark' account of quality of life as they provide no account of pleasure or enjoyment of life (Phillips 2006: 90). Furthermore, the role of participation in defining needs-based approaches in general has been unclear (Alkire 2002). In addition, critics of needs-based approaches generally claim that they are overly prescriptive and concentrate on resources (a bundle of goods) and functionings (doings and beings) rather than the capability to access resources or carry out functionings. However, there are considerable overlaps between the capability and needs-based approaches. O'Neill (2011: 28) argues that using verbs rather than nouns to describe needs, 'Joe needs to be able to eat adequately' rather than 'Joe needs food', shows that the capabilities and needs approaches may not be as far apart as often thought. Gough (2003:16) himself admits that, 'the functioning–capability distinction would help us to diminish lingering charges of paternalism'.

Summary

The contribution of these philosophical endeavours was to refine important questions about the aim of public policy. Should it aim to increase an aggregate measure of happiness in society or for equality? If the latter, what sort of equality: of happiness, resources, capability or fulfilment of needs? This brief account cannot fully represent the extensive, complex and often highly abstract debates in the literature about the proper focus for wellbeing measurement. However, the philosophical framework outlined here may help us understand the particular theoretical or ideological stance on wellbeing policy that public bodies may be considering, and the types of prescriptions and solutions proposed in policy discourses. Sabina Alkire (2002: 194) reviewed a range of lists setting out dimensions of human development, reflecting a wide range of theories about wellbeing and quality of life. She pragmatically concluded that what was most important about a list was its efficacy as a means to 'confront the many challenges of this generation'. So, although theoretical clarity can enable better informed policy decisions and enhance the link between indicators and action, the search for philosophical perfection may distract.

Happiness and hard science – subjective wellbeing

In addition to these theories, work in economics and psychology over the last 40 years has created a large body of statistical evidence about subjective wellbeing (SWB). This is a complex field that generates a great deal of interesting debate. Essentially SWB research focuses on the individual's own assessment of their wellbeing. Subjective wellbeing can be crudely categorised in two ways: hedonic and eudaimonic (Sirgy *et al.* 2006). Hedonic wellbeing research focuses on measuring 'happiness' (positive affect)[10] or judgements about life satisfaction (cognitive evaluation). Measurements are usually taken using large-scale surveys that ask people to rate their feelings (affect) or life satisfaction judgements on a Likert scale. These are then compared with other variables to tell us, for example, whether there is a positive correlation between life satisfaction and income. Life satisfaction surveys can also be split into 'domains' such as work, relationships, leisure, environment, et cetera, so measurements can tell us how different variables correlate to reported satisfaction in different aspects of life. For example, a higher income may correlate with satisfaction in work but not relationships. Life satisfaction studies have long been used by economists, and the focus on the individual's judgement about their own welfare according to their own criteria fits well into a utilitarian framework. In addition, individuals' assessments of their subjective wellbeing have been shown to be reliable when compared with external reports (of friends or family) and to observed behaviour (Helliwell and Putnam 2004; Diener 2000). Recent advances in neuroscience have now located the site of activity in the brain that corresponds to happiness, thereby bringing happiness research into the realm of hard science (Layard 2006). Scientists have found that when people report on their own levels of happiness, this largely corresponds to an 'objective reality', a measure of activity in the brain that rises and falls 'just like your blood pressure' (Layard 2003: 8). This measurable phenomenon has made SWB more interesting to economists, who are realising that positive affect may after all be quantified.

Eudaimonic wellbeing is a more recent and fast-developing field of research, which recognises the limitation of single-item measures of happiness and life satisfaction studies and focuses on a fuller conception of wellbeing, how well people function and how they realise their potential (Huppert *et al.* 2009). Work in this field has been influenced by Aristotelian ideas of human 'flourishing' and the capabilities approach and has been informed by the work of psychologists such as Carol Ryff who developed an approach identifying six key components of wellbeing: autonomy, environmental mastery, personal growth, positive relationships, purpose in life and self-acceptance (Ryff 1989). Other psychologists have developed a psychological needs-based approach to SWB. The self-determination theory (Ryan and Deci 2000) defines autonomy, competence and relatedness as three psychological requirements for wellbeing. This resonates with the needs approach discussed earlier where autonomy and health (arguably 'competence') are primary needs defined by Doyal and Gough's *Theory of Human Need*. While many use the terms hedonic and eudaimonic to distinguish between different approaches to subjective wellbeing, some consider the latter approach as a distinct but complementary field to SWB research termed psychological wellbeing, a more holistic and 'theory-guided approach to well being' (Forgeard *et al.* 2011: 93–4; Sirgy *et al.* 2006; Ryff and Keyes 1995). These terminological differences are indicative of a fast-moving and emerging field.

GDP and subjective wellbeing – the Easterlin paradox

An iconic and long-debated finding of SWB research is that in advanced liberal democracies in recent decades average life satisfaction has remained flat despite considerable increases in GDP, the so-called 'Easterlin paradox' (Easterlin 1974; for a discussion see Jackson 2009; Jordan 2008; Offer 2000; Johns and Ormerod 2007). Many subsequent studies show that SWB and economic growth are correlated but only up to a certain point, after which there seems to be little relationship: more and more money does not increase happiness. Furthermore, the sum at which income stops influencing SWB is surprisingly low. Gasper (2007b: 486) describes this as 'one of the major findings of modern social science' and it has become a mantra in a range of alternative contemporary discourses whose common aim is to discredit mainstream economic growth policies.[11] However, the paradox theory is not without its critics (Offer 2000). As Ormerod (2007) suggests, people can choose the evidence to fit their argument. There is a tendency for SWB data to be uncritically 'grafted onto' existing concepts such as capabilities or environmentalism to back up their various claims (Jordan 2008: 15). Whilst the Easterlin paradox may have its detractors, it is clear that a more complex relationship between income and happiness exists than the one that informs utilitarian economics.

Studies have consistently shown that, after a basic income threshold, relative income has a greater impact on happiness than absolute income. So even if the income of the poorest is rising, SWB will not necessarily be increased if the richest are also getting richer. The study of 'gaps', or multiple discrepancy theory, puts the case that out of a series of gaps between what a person has and wants; has and needs; has and relevant others have; social comparisons have the most effect on satisfaction and happiness (Michalos 1999). There is additional evidence that the reference area for social comparison is more likely to be national than local, which means that social position in wider society has more effect than direct comparison with neighbours (Ballas *et al.* 2007; Wilkinson and Pickett 2006). This is an unpalatable message for the British government because, although general affluence is increasing, the gap between the richest and poorest in UK society is also growing, and certain sections of society are being left further and further behind (Jones 2008; Ballas *et al.* 2007; Marmot *et al.* 2010). Wilkinson and Pickett (2010: 174) present statistics from 21 countries in developed market democracies to show that health and social problems are higher in more unequal societies (using income inequalities) and that the UK is the third from the bottom of that list. 'Our growing understanding of how human health and wellbeing are so deeply affected by social structure inevitably pushes science into politics' (Wilkinson and Pickett 2010: 277).

This body of SWB research incorporates studies which explore the relationship between materialism and SWB and report a negative correlation. Those people who have predominantly 'intrinsic' values, focusing on building loving relationships and self-development are happier than those who hold 'extrinsic' values focusing on status, material wealth and image (Schmuck *et al.* 2004; Kasser and Ryan 1996). Therefore, in addition to the evidence that GDP is not an adequate measure of wellbeing and that equality is important, there is a growing debate about whether policies focused on economic growth and consumption, fostering a rise in an image-dominated materialistic culture, are actually reducing SWB. This has become a powerful discourse promoted in a number of books written for the popular press, such as *Affluenza*[12] by Oliver James (2005), and organisations, such as the New Economics Foundation which, in its *Wellbeing Manifesto for a Flourishing Society* (Shah and Marks 2004), called for a ban on commercial advertising pitched at children.

A health warning

In recent years, the amount of research on SWB has 'burgeoned' but there are all sorts of problems in measuring SWB, including a lack of evidence on causality, making clear policy recommendations problematic (Dolan *et al.* 2008: 96). Despite increasing attempts to theorise SWB and some claims for 'wellbeing theory' (Forgeard *et al.* 2011), the SWB research field is an 'emerging science' and there is a lack of consensus about what should constitute a 'gold standard' of wellbeing measurement (Huppert *et al.* 2009: 305). Subjective wellbeing should be treated as a 'broad area of research activity and interest, rather than as a specific construct' (Sirgy *et al.* 2006: 388). It is characterised by a proliferation of discrete studies, which tend to be cross-sectional rather than longitudinal, each using a different categorisation of variables and choice of categories (for example, some studies on the effect of long-term relationships treat married and cohabiting people separately, while others group them together) making universal claims problematic (Jordan 2008; Dolan *et al.* 2008, 2006b; Sirgy *et al.* 2006; Diener and Seligman 2004). In addition, there are various problems with relying on people's own assessment of their wellbeing because SWB can be adaptive or distorted (see Gasper 2007b for a discussion). Nevertheless, some clear correlations and patterns have emerged between SWB and relative income, age, health, personality, relationships and employment status (see Diener *et al.* 2009 for a review of evidence). This new body of data, and increasingly sophisticated measurement tools, have allowed scholars to interrogate how subjective wellbeing relates to different personal, societal and (increasingly) political variables. The field is growing fast and often provides interesting evidence but, as Flavin *et al.* (2011: 253, f1) warn, scholars should be 'careful not to attribute explanatory power' to surveys 'beyond what they represent'. Despite earlier complaints that SWB was largely ignored by policy actors in favour of objective indicators, international policy discourses have become increasingly infused with references to wellbeing and happiness and many governments are now creating national subjective wellbeing indices (Bache and Reardon 2011). Although, considering the problems attributing causality, it remains largely unclear how these indices will influence policy decisions other than to serve as an important 'cross-check on the validity of economic and other social indicators' (Sirgy *et al.* 2006).

Societal wellbeing, social quality and social capital

The discussion so far has centred mainly on different models of individual wellbeing, whether this be the preference-based utility model, approaches based on capabilities or needs, or the research on subjective wellbeing. Many scholars defend ethical individualism as the appropriate approach for wellbeing definition and measurement as it allows for the fact that there are important differences between people and that social relations and culture can exert harmful constraints upon individuals (Robeyns 2003). In addition, some theorists see an approach focused on meeting individual needs as the basis for promoting social citizenship. As Gough states, 'It is inconsistent for a social group to lay responsibilities on some person without ensuring she has the wherewithal to discharge those responsibilities' (2003: 5). Such accounts make obvious claims to human rights and civil liberties for each and every person. However, such accounts are also critiqued for tending to view human relationships and social structures as instrumental to individual wellbeing instead of having intrinsic value; they may set up a false polarisation of the 'I' and the 'we' instead of recognising their inherent co-construction (Oughton and Wheelock 2006). On this basis, scholars argue for institutional approaches to wellbeing that look to integration of state, market, organisation, community, household and individual (Prilleltensky and Prilleltensky 2007; Oughton and Wheelock 2006). Similarly public health scholars and social policy theorists point to a lack of focus on the importance of culture in influencing wellbeing (Jordan 2008; Eckersley 2006, Carlisle and Hanlon 2007).

Jordan (2008) argues that what matters is 'how institutions and cultures enable individuals to value their experiences, rather than simply what they have the opportunity to do' calling for a focus on 'social conventions and power processes' as these ultimately 'relate behaviour to wellbeing' (Jordan 2008: 25–26). He critiques the capabilities approach on this basis, and Nussbaum in particular. This is echoed by Michalos (2011: 121) who points out that focusing only on the freedom of what individuals *can* be and *can* do is 'troublesome' in the sense that it provides no guidance as to what is either moral or pleasant in life. He points out that one could be 'fully free and able to pull a drowning child from the water but choose instead to continue sunbathing on the beach'. While I have instinctive sympathy with the need to consider social values and institutions, as will be evident from this book, Jordan's criticism is unfairly directed at Nussbaum's account, which refers throughout to the importance of protecting social cultures and institutional arrangements that help people flourish. For example, she talks of the 'social bases of self-respect' and 'protecting institutions that constitute and nourish [empathetic and meaningful social affiliations]' (Nussbaum 2000: 79). As Phillips (2006: 183) argues the capabilities approach 'has considerable resonance with the definition of social quality' based heavily as it is on participation. Nussbaum has been criticised elsewhere for *failing* to distinguish between individual needs and societal conditions (Gough 2003: 14). Nussbaum's list is included in several of her texts in which she clearly acknowledges the role of social culture and institutions, for good or ill, in affecting individual wellbeing.

These discussions are hampered by a lack of clarity regarding terminology. 'Social wellbeing', which is a characteristic of individuals referring to the quality of their social relationships, should be distinguished from 'societal wellbeing', which refers to the quality of society. However, some authors, for example Phillips (2006), categorise the capability approach as an example of a 'social perspective' of wellbeing due to its focus on social justice and the social and political basis for this; whereas others, like Jordan (2008), seek to distinguish it from an account based on 'social value'. The UK Office of National Statistics has a programme of work to measure wellbeing. Measuring 'subjective wellbeing' is a sub-section under 'Measuring Societal Wellbeing'. When looking on their website for a definition of societal wellbeing the closest I was able to find was:

> 'Measuring Societal Wellbeing' provides an overview of measuring societal wellbeing and considers the main approaches emerging for how it should be measured…There is no shortage of relevant existing data to help build pictures of societal wellbeing and to assess our quality of life. Clearly, our overall wellbeing is likely to reflect health, education, culture, safety and a sense of community, among other things. But any of the strands may be valued more or less by individuals and groups within society.
>
> (ONS 2011)

Is societal wellbeing simply an aggregate of individual wellbeings or something qualitatively different? If the latter, then what is it? Theories of social quality have been thoroughly developed by Beck *et al.* (1997; 2001a) and emerged from a critique of the EU's focus on an economic development paradigm. In 1997 the Amsterdam Declaration was signed by 74 academics in the social sciences and the EU has now incorporated social quality into its social reporting (see Phillips 2006 for an extensive review). According to Phillips (2006: 176): 'The social quality of a collective is not just the accumulation of the life quality of each of its individual members: it incorporates collective as well as individual attributes and is holistic in its orientation.' The major components which make up social quality are socio-economic security, social cohesion, social inclusion and social empowerment (Beck *et al.* 2001b: 352). This construct is measured in the European Network on Indicators of Social Quality (ENIQ) and includes 95 indicators under the sub-headings of (for example): income security; security of health provisions; working conditions; generalised trust; altruism; social contract; political rights; social care; friendships; availability of information; reconciliation of work and family life; cultural enrichment; support for collective action; openness of economic system; support for social interaction. These indicators perhaps get closer to what Jordan (2008) is talking about when he refers to 'social value'. However, these societal level theories are not well referenced in current wellbeing research literature which is dominated by economic and psychological approaches to subjective wellbeing measurement. Rather, SWB studies have forged a closer link with the social capital literature which tends to focus more narrowly on civic associative behaviour and relationship networks, a more social than societal wellbeing construct. In a review of wellbeing research literature Cronin de Chavez *et al.* (2005: 77) found that most research is focussed on physical and psychological aspects of wellbeing and that the 'social and cultural bases of wellbeing' have been under-explored. They found large disciplinary gaps and a surprising lack of research in the sociology and anthropology fields.

This is well illustrated by a recent attempt 'to go beyond individualistic aspects of wellbeing, by incorporating measures of social or interpersonal wellbeing' (Huppert *et al.* 2009: 304). The theory they refer to is social capital and questions in this domain are: the levels of help and trust in the local area, quality of family life and relationships, levels of giving and participating in local activities, optimism about the future and whether the individual feels they get enough recognition and are treated with respect and fairness. However, these questions are still trying to measure what is important for the individual rather than what constitutes societal wellbeing. A range of SWB studies show that social relationships are extremely important for individual wellbeing and that volunteering and community participation are correlated with higher levels of SWB (Dolan *et al.* 2008). Helliwell and Putnam (2004) found social capital and SWB to be positively correlated (although they guard against assumptions about causality) across areas such as family and social life and community involvement. However, someone may have a good score on all these factors but have extremist views and volunteer for a terrorist organisation.

It is often easier to diagnose that something is lacking than to define what this actually is. Many point to the 'absence of society' and synonymise this with a decline in societal values, and increasing negotiations by individuals regarding choice and risk linked to market mechanisms (Bauman 2008; Beck 1992). It is now a common refrain that in western liberal democracies a weakening of traditional ties, moral relativism and increased individualism and materialism are the cause of social decline. The inference is that the breakdown of families and high levels of insecurity, distrust and violent crime can be solved by having more of 'society'. This is an over simplification and conflation of a set of complex phenomena (Grenier and Wright 2006). A perceived decline in social values since the 1960s is often linked to the increased freedoms that era brought for particular social groups. This sense may be fuelled by nostalgia for the 1950s and 1960s 'a comparatively crime free era of strong communities and solid families (albeit with endemic homophobia, sexism, poverty and racism)' (Browne 2008: 4). This association of entities is common. Its inference is that the civil rights and liberties movements of the 1960s and 1970s brought increased crime, community collapse and family breakdown. In a complex period of social transition set in a globalised context of massive change it is by no means a given that, for example, a greater focus on women's rights will lead to a reduction in family or social values, even though this may have contributed to a rise in divorce. There are many reasons for seeking divorce, but common discursive inferences will affect how we view the definition, causes and therefore solutions of social problems.

These discourses of social decline infuse and in many cases are the basis for theories of social capital. There is now a very wide literature on social capital (for a thorough and accessible review see Field 2008). Like 'wellbeing', social capital is a term that has been gaining currency, particularly since the publication in 2000 of Robert Putnam's influential book *Bowling Alone*. Definitions of social capital include 'networks of sociability, both formal and informal' (Hall 1999: 418); 'social networks and the norms of reciprocity and trustworthiness that arise from them' (Putnam 2000: 19); or simply, 'relationships matter' (Field 2008: 1). Putnam argues that social capital can be both a private and a public good, and that social networks can have externalities which affect general wellbeing levels, not just those of the participants. Putnam also stresses that like any other form of capital (e.g. physical or economic) social capital can be used for good or ill. Strong ties between individuals can promote sectarianism and corruption just as they can promote co-operation and trust (2000: 22). Putnam makes a distinction between social capital that bonds and social capital that bridges. The former refers to bonds based on specific reciprocity that tend to occur within tight-knit groups. The latter refers to the links between different networks of people. The first is vital social 'glue' and the second is 'lubricant' needed for social mobility. Jordan (2008) critiques the language and claims of social capital on the basis that it promotes very little that cannot already be explained through economic models of increasing individual utility, and that Putnam even uses the language of economics to frame the concept. Grenier and Wright (2006) argue that social capital theory and research has concentrated too much on face-to-face, non-market, associational and volunteering activities and ignores important realms such as the workplace or internet and phenomena like informal childcare networks. They find that although overall levels of associational activity and charitable giving have stayed stable in the UK, more (time and money) is given by fewer people today in different ways and for different reasons than previously. In short, they find that participation is more concentrated in the A, B and C1 social classes and is becoming increasingly commodified, for example, volunteering in order to enhance a CV. They argue that participation is becoming more and more 'about private benefit, hollowing out its social meaning and weakening its relationship with social capital' (2006: 48). In Jordan's (2008) view, someone who volunteers may do so for all sorts of reasons and it is these *reasons* which are important in assessing social value, and it is social value which creates the link to wellbeing.

Discourses of wellbeing and public policy in the UK

In recent years wellbeing has risen up the political agenda in the UK to the extent that some commentators view it as 'an idea whose time has come in British Politics' (Bache and Reardon 2011: 1). In 2002, a 'life satisfaction seminar' was held at the Treasury where economist Professor Richard Layard (also a member of the House of Lords and nicknamed the Happiness Czar by some pundits) urged government to rethink its priorities and promote the happiness of people (Easton 2006). This generated a modest interest in exploring the link between government activity and life satisfaction, as well as wellbeing in general (Donovan and Halpern 2002). The Department for Environment, Food and Rural Affairs (Defra or 'Happy Central') was charged with developing a measure of wellbeing (Furedi 2010; Easton 2006) and commissioned several reviews of wellbeing research and its policy implications (such as Dolan *et al.* 2006b; Levett-Therivel Sustainability Consultants 2007). Quality of life indicators had already been developed by Defra under the sustainable development paradigm (and these are discussed in the next chapter) but in 2005 the new UK National Sustainable Development Strategy *Securing the Future* stated a stronger commitment to explore the concept of wellbeing. Although wellbeing is not defined it is mentioned alongside 'life satisfaction' and 'quality of life'. The document also states that 'the issue of wellbeing lies at the heart of sustainable development, and it remains important to develop appropriate wellbeing indicators' (Defra 2005: 23). This suggested that something new, other than quality of life which was already being measured in SD indicators, was being explored here.

The Whitehall Wellbeing Working Group was set up in 2006 and developed a statement of common understanding regarding the definition of wellbeing:

> Wellbeing is a positive physical, social and mental state. It is not just the absence of pain, discomfort and incapacity. It arises not only from the actions of individuals, but from a host of collective goods and relationships with other people. It requires that basic needs are met, that individuals have a sense of purpose, and that they feel able to achieve important personal goals and participate in society. It is enhanced by conditions that include supportive personal relationships, involvement in empowered communities, good health, financial security, rewarding employment, and a healthy and attractive environment.
>
> Government's role is to enable people to have fair access now and in the future to the social, economic and environmental resources needed to achieve wellbeing. An understanding of the combined effect of policies and the way people experience their lives is important for designing and prioritising them.
>
> (Defra 2007)

In particular, work was led by the Young Foundation, in conjunction with Professor Layard and the Improvement and Development Agency (IDeA) on developing local wellbeing indicators in a pilot project with three local authorities (South Tyneside, Hertfordshire and Manchester) focusing on interventions to reduce depression and social isolation and increase emotional resilience (Mguni and Bacon 2010; Hothi *et al.* 2008; Steuer and Marks 2008). This work has been heavily influenced by the work of American psychologist and founder of the 'Positive Psychology Movement' Martin Seligman and in the pilot areas cognitive behavioural therapy techniques were being taught to all 11- to 13-year-old students in schools. The UK Resilience programme received much positive and negative attention in the media at the time, major criticisms being that this was yet another sticking plaster policy because it focuses on the individual's personal ability to control their emotions rather than looking at the effect that current political economy has on society (Furedi 2008). The role of think-tanks has been very influential in the UK. The New Economics Foundation, for example, an independent think-tank based in London, has long spearheaded the campaign for quality of life, sustainability and (more recently) wellbeing to be a central part of public policy. They have played a key role in lobbying for and developing innovative local and national, individual and community, quality of life and wellbeing measurements and indicators (Marks *et al.* 2004).

Discourses of wellbeing in the UK are shifting and have a 'holographic' quality (Ereaut and Whiting 2008: 5) meaning different things are projected onto them. Increasingly, the three disciplines of economics, psychology and neuroscience merged into what Carlisle and Hanlon (2007) term the 'scientific discourse' of wellbeing promoted by several high-profile wellbeing gurus, including economist Richard Layard, positive psychologist Martin Seligman and Nic Marks from NEF. A whole series of books emerged in the mid-2000s which review and discuss the nature of this new science (for example, *The Science Behind your Smile,* Nettle 2006; *Happiness: Lessons from a New Science,* Layard 2006; *The Science of Wellbeing,* Huppert *et al.* 2005). Many conferences followed, where broadly the same group of keynote speakers did the rounds. In the UK media there were regular articles and debates about happiness, an explosion of books on the subject, and several television programmes explored the concept including the BBC's *Making Slough Happy* in 2005 and *The Happiness Formula* in 2006. This prompted discussion around whether measuring and promoting subjective wellbeing should be a goal of public policy. Although researchers have always been keen to distinguish 'happiness' from SWB as it is an 'unwieldy construct for scientific research' (Forgeard *et al.* 2011), it nevertheless was and still is a central notion in these discourses.

In a discussion paper on future strategic challenges for Britain from the Cabinet Office Strategy Unit it stated:

> With four in five Britons believing that the Government's prime objective should be the greatest happiness rather than the greatest wealth, future politics will need to deal both with the 'bread and butter' political issues and to address issues likely to affect citizens' wellbeing…over the next ten years government will be increasingly measured against how happy it makes its citizens.
>
> (UK Government Cabinet Office 2008: 173)

The creation of an All Party Parliamentary Group on Wellbeing Economics in March 2009 was led by a Liberal Democrat MP. These initiatives were part of a wider movement in Europe and beyond including the 2007 OECD 'Beyond GDP' international conference (see Abdallah *et al.* 2011 for a review). In 2008, Nicolas Sarkozy set up the Commission on the Measurement of Economic Performance and Social Progress chaired by Stiglitz, Sen and Fitoussi (2009). This was influential on the UK wellbeing effort and in November 2010 the Coalition Government announced a strategy to measure national wellbeing and charged the Office of National Statistics to carry out a public consultation with the aim of designing a set of wellbeing survey questions that it could incorporate into its national survey:

> It will open up a national debate about what really matters, not just in government but amongst people who influence our lives: in the media; in business; the people who develop the products we use, who build the towns we live in, who shape the culture we enjoy.
>
> (Cameron 25 November 2010)

There is now a website promoted by the Young Foundation called Action for Happiness (formerly Movement for Happiness) with many of the key players (Richard Layard, Nic Marks, Geoff Mulgan, Anthony Seldon) involved. They explicitly link their theory of happiness with Bentham's under the tab '*the* philosophy of happiness' [my emphasis]. In doing so they are espousing a 'new utilitarianism' (Jordan 2008). They promote 'ten keys to happier living' including giving, noticing, and relating (to others) and this is heavily framed within the context of social and community relations and volunteering.

Some argue that the new discourses of wellbeing promote a reductionist view of wellbeing and focus attention away from the social and political basis of wellbeing onto an individual model where people are responsible for their own wellbeing, recasting social problems as the problem of the individual (Jordan 2008; Edwards and Imrie 2008; Furedi 2008). Edwards and Imrie (2008) scrutinised these discourses in the context of disability policy in the UK. They suggest that by promoting a self-actualisation view of wellbeing, they do little to contribute to the understanding of disablement in society; indeed they signal a 'retrograde step' from the concerted attempts of lobby groups to 'shift interpretations of disability from individualised, biological, conceptions based on internal limitations, to ones situated in the socio-structural relations of an ablist society' (2008: 338). They argue for 'collective politics' regarding the definition of problem and solution at a societal level rather than at the level of the individual. They are among a number of critics who claim that in current wellbeing measurement discourses and practice 'far too little attention has been devoted to theorizing about how socio-political conditions determine quality of life' (Flavin *et al.* 2011: 265; Tsai 2011).

Definitions, domains and dimensions

Although I started this chapter talking about wellbeing and quality of life as interchangeable, in the above review we can see how wellbeing discourses have changed and how the nature of these different discourses are linked to wider agendas and themes like social capital. Although Galloway (2005) notes that 'Some regard the terms as interchangeable, while others regard wellbeing as one component of a broader concept of QoL', I think Gasper is broadly correct regarding public discourse when he states:

> The 'well-being' (WB) term is used more when we speak at the level of individuals, and quality of life (QoL) somewhat more when we speak of communities, localities, and societies. Similarly, 'well-being' is used somewhat more to refer to actual experience, and 'quality of life' more to refer to context and environments.
>
> (Gasper 2009: 5)

Gasper views both as overlapping concepts whereas 'wellbeing' is predominant in psychology and 'quality of life' came from social policy. In economics and psychology fields 'wellbeing' tends to refer to subjective wellbeing, self-

Table 2.1 Quality of life domains

Core QOL domain	Indicators
Emotional wellbeing	Contentment
	Self-concept
	Lack of stress
Interpersonal relations	Interactions
	Relationships
	Supports
Material wellbeing	Financial status
	Employment
	Housing
Personal development	Education
	Personal competence
	Performance
Physical wellbeing	Health
	Activities of daily living
	Leisure
Self-determination	Autonomy/personal control
	Goals and personal values
	Choices
Social inclusion	Community integration and participation
	Community roles
	Social supports
Rights	Human
	Legal

Source: Schalock and Verdugo (2002) cited in Galloway (2005: 26).

assessed happiness, welfare or utility. However, an increasing number of scholars distinguish 'wellbeing' from 'welfare' to signal a move away from the idea of utility maximisation (Jordan 2008; Canoy and Lerais 2007). Quality of life, on the other hand, is often regarded as a concept used to construct *objective* indices, for example in comparing different places to live. Some reject the objectivity of such indices as they feel that they 'simply reflect the values of those who construct them' (Bell 2005: 98).

Galloway (2005) found that 'quality of life' is most often used to identify a multi-dimensional concept which is made up of a number of 'domains' such as health, education and social relationships. In a major review of academic quality of life studies, Schalock and Verdugo (2002) compiled a list of the most common quality of life domains and descriptors extracted from 2,455 academic articles (see Table 2.1).

Although the number and range of domains varies, there is considerable overlap, as well as significant cultural differences, in the core components of quality of life and therefore there are problems with producing a definitive list of what constitutes it. Rapley, like Galloway, argues that 'instances of the usage share a family resemblance but they do not point to a singular, underlying 'thing' or 'entity' and 'despite the immense outpouring of QoL studies since the 1960s we are no closer to knowing how to answer questions such as "what is quality of life" ' (Rapley 2003: 215). Perhaps the most that can be claimed for quality of life is that it is 'a multi-attribute value concept' which is 'subject defined and subject dependent' (Vlek *et al.* 1998: 321), and which brings together various dimensions that are considered important for the goodness of life. Nevertheless, rather than 'hang it up as a hopeless term' it is useful in helping us to frame discussions:

> I have come to believe that the concept of quality of life…may offer not so much a formalised, psychometric, conceptual framework for understanding quality of life as a human universal. Rather quality of life may offer us a sensitizing concept for thinking through the purpose and methods of delivery of human services, or ways to enhance the 'liveability' of our particular communities in a democratic, inclusive and emancipatory way.
>
> (Rapley 2003: 212)

Similarly with regard to wellbeing, MacGillivray (2007a) identifies common wellbeing dimensions as knowledge, friendship, expression, affiliation, bodily integrity, health, economic security, freedom, affection, wealth and leisure. Dinham (2006: 183) notes the absence of consensus about the meaning of wellbeing:

> Thus there emerges a range of factors identified as inherent in it or against which it is recognizable and/or measurable. Yet, at the same time, there is little or no consensus about what it really means or looks like and therefore to produce and reproduce it, and to know that it is there, proves highly difficult except in the most general of terms.

Yet he argues that the concept of wellbeing is useful as it allows us to try to get a 'glimpse of what is important'.

Summary

In the last half century our understanding of wellbeing and quality of life has been broadened outside the ambit of economic growth. A variety of different theories of quality of life and wellbeing have emerged, and attempts at application have produced important lessons for the measurement of social phenomena. The dominant theory of welfare based on economic growth was challenged in the 1960s. A proliferation of alternative social indicators reflected a new awareness that the concept of quality of life was more complex. However, the profusion of measurement, the lack of theoretical clarity and the changing social and political culture hampered the effectiveness of the social indicators movement to influence policy. Ideas of wellbeing were influenced by neo-liberal discourses that promoted quality of life as something belonging to autonomous individuals which could be enhanced in the market place. These discourses rested on normative values of freedom, liberty and rights. In addition, ideological tensions arose between quality of life as a societal or individual construct and between objective and subjective measurement. Although the social indicators movement did not have the sort of policy influence that its advocates hoped, it helped to generate an immense amount of new thinking and research into defining and measuring wellbeing. More complex notions of wellbeing focused, for instance, on equality and capabilities to flourish, have been influential in producing quality of life indices, measuring factors like life expectancy, health, education and civic participation.

A new 'second wave' of wellbeing driven by discontentment with GDP and increasing environmental concerns is sweeping across UK politics and beyond (Bache and Reardon 2011: 6). Growing concern amongst many that the way our political economies are organised around production and consumption is producing diminishing returns for our quality of life and failing to decrease inequality or environmental damage. A large body of academic work in the SWB measurement field has amassed evidence to show a complex relationship between income/GDP and SWB. This has been forcefully mobilised in popular and policy discourses about happiness in the UK. These same discourses have come under criticism for their over-reliance on an individual construct of wellbeing and a lack of attention to social culture and political structures and processes. The next chapter will consider the theory of sustainable development and its relationship with, and impact upon, human wellbeing theory and measurement.

3   Sustainable wellbeing: an oxymoron?

A few years ago, I was part of a British Council academic exchange with researchers from Obafemi Awolowo University in Ife-Ife, Nigeria. We were working on a sustainable water management project which involved visiting rural communities. We spent a lot of time in a very hot minibus. On one particular journey someone decided it would be a good idea to swap jokes and I was expected to come up with British jokes in exchange for Nigerian ones. Since I only know one joke (which they didn't get) and as I didn't get any of their jokes, this was a complete failure. Later, after having a few drinks with everyone at the local bar and displaying impeccably professional behaviour, I fell into a concrete drainage ditch and gashed my leg. While waiting with me in the local medical centre, one persistent member of the team told me this 'sustainability joke' to cheer me up. To my delight this was the one Nigerian joke I got:

> One day a farmer walked to the local town. As he entered the main street he noticed a new chop house and was taken aback by the sign in the window which read, 'Come in friends! Eat all you want! Your grandfather has already paid!' He moved closer in disbelief and read the sign again, at which point the door opened and the proprietor invited him inside:
> – Come in my friend! Eat all you want! Your grandfather has already paid!
> – Are you sure? My grandfather?
> – Yes my friend, come in, come in, and eat your fill!
> Well, who can resist free food? The farmer entered and sat down and ordered a whole goat's head, a pile of pounded yam, some okra and chilli. He stuffed himself to the gills and then ordered the whole thing again and stuffed his pockets for later. As he got up and staggered to the door the proprietor came running after him:
> – My friend, my friend, where are you going? You have not paid!
> – But you said my grandfather has already paid!
> – Yes, yes, my friend, your grandfather has already paid for you, but now it is your turn to pay for your grandchildren!

Concern about the wellbeing of future generations is a central tenet of sustainable development theory. Yet as we have seen in the last chapter there is much debate about how we can know and support the wellbeing needs of ourselves and our fellow humans let alone those yet to come. Phillips (2006: 244) describes this as 'one of the most pressing long term issues the world is facing in the twenty-first century'. This chapter considers the extra theoretical and practical challenges that a concept of sustainability poses for wellbeing definition and measurement.

Despite the developments explored in the last chapter, and the fact that neoliberalism struggles to retain any last vestiges of political credibility, economic growth still underpins dominant understandings of human development. It has long been seen as the key determinant of human wellbeing, securing greater access to healthcare, education, employment and autonomy for citizens (Marks *et al.* 2006b). However, unchecked production and consumption linked to economic growth has produced global environmental problems. These problems, such as climate change, pollution, natural resource depletion and reduction in biodiversity threaten the stability of the planet itself. For example, at current best estimates, a 50–85 per cent reduction in CO_2 emissions by 2050 is needed to address the risk of dangerous climate change.[1] With the continuing rise in the global population and growth of emerging new economies, this is not feasible. International organisations and governments worldwide are re-assessing agendas for economic growth.[2] In future, these must be linked more closely with environmental wellbeing objectives in order to avoid hardship in the long term.[3] Over the last 40 years, concerns about the biophysical integrity of the planet and the impact of environmental problems on the poor have seen powerful ecological and environmental justice discourses develop alongside philosophical debates about our relationship with nature. The emergence and nature of these environmental discourses is fully and excellently explored elsewhere (for example: Dresner 2002; Dryzek 1997; Macnaghten and Urry 1998; Lash *et al.* 1996). This chapter will concentrate on the concept of sustainable development 'arguably the dominant global discourse of environmental concern' (Dryzek 1997: 123) and its relationship with human wellbeing. This widens and complicates the notion and measurement of quality of life by seeking to promote environmental quality and the wellbeing of future generations as key factors. The two fields of human wellbeing and sustainable development (SD) measurement have, to a large extent, developed separately (Gasper 2007a; Neumayer 2007) although there have been recent endeavours to integrate the concepts (for example Marks *et al.* 2006b; Dolan *et al.* 2006b; Stiglitz *et al.* 2009; Rauschmayer *et al.* 2011). This chapter set out the relationship between concepts, focusing particularly on discourses in the UK and local indicator development.

Background

Sustainable development as a concept has been around since the 1970s but gained momentum in international policy terms in 1987 when the popularised Brundtland version of SD was produced. It arose from the final report of the World Commission on the Environment and Development (WCED) set up by the UN General Assembly in 1983 to make the idea of sustainable development politically acceptable and credible. Previously, poorer countries had tended to frame environmental concerns as a problem of richer nations. Prime Minister Indira Ghandi's sentiment that 'Poverty is the worst form of pollution' was symptomatic of the polarisation in environmental attitudes between North and South (Najam 2005). The World Conservation Strategy in 1980 foreshadowed many of the Brundtland Commission's recommendations, including a definition of SD. However, it was perceived as being written predominantly by northern environmentalists and its focus on the environment was not popular with development agencies (Dresner 2002). It also included no account of the political or economic changes seen as necessary. The Brundtland report, which produced the oft-quoted definition of SD[4] sought to make links between environmental degradation and poverty in a way that could bring both rich and poor nations to a shared understanding. This was apparent in the choice of the report's title, *Our Common Future*. It sought a way forward that was acceptable to environmentalists and to business interests in the north (Boulanger 2007). It employed a basic needs and social justice account of human wellbeing, reflecting its desire to enlist developing nations, while being cautious in its wording on the need to limit consumption, a potentially unpopular message to bodies with business interests. SD has had to appeal to diverse and conflicting interests from the start, which is both its weakness and its strength. In the west, neoliberal economic discourses of production and consumption dominate concepts of human wellbeing and the same is true for SD. However, although dominant economic interests, structures and incentive systems still hold sway, as a way of conceptualising a different development path SD has gained enormous legitimacy.

A distinctive tenet concerned safeguarding the basic needs not only of present generations but also those in the future, and this underpinned the notion of environmental sustainability in the report. The timing of its release during a wave of global anxiety about the environment, following the discovery of a hole in the ozone layer in 1985 and the Chernobyl disaster in 1986, helped its message gain international currency (Dresner 2002). The key features of the approach are set out in Box 3.1.

The 1992 United Nations Conference on Environment and Development (UNCED), better known as the Rio Earth Summit, took the Brundtland report further and sought to instigate an action plan. This emerged as Agenda 21, a lengthy and complex document, agreed by 171 nations. Agenda 21 focused on implementation of SD and particularly stressed the importance of bottom-up participation through Local Agenda 21 (LA21), a community-based approach focused on citizen participation, education and action with the rallying cry 'Think global, act local'.

Sustainable development and sustainability

SD and sustainability tend to be used interchangeably. Sustainability is a much older concept used in ecology referring to 'the potential of an ecosystem to exist

Box 3.1 Summary of the Brundtland approach (after Baker 2006).

Sustainable development

- Acceptance of biophysical limits to growth
- Recognition of the link between environmental problems and poverty
- Recognition of the unsustainable consumption patterns of the North
- Argument for a win-win between environment and development
- Development focused on basic needs, alleviating poverty and social justice
- Constructs the environment/economy/equity model of development
- Calls for new governance structures for environmental management from local to global level

Key normative principles:

- Common but differentiated responsibility
- Inter- and intra-generational equity
- Justice
- Participation
- Gender equality

over time'; adding the concept of 'development' shifted the focus from ecology to society (Baker 2006 citing Reboratti 1999). Previously sustainability was predominantly a scientific concept originating in natural resource management (Bell and Morse 2003). Dresner (2002) argues that despite the obvious logic that sustainability and SD must be different (otherwise the word 'development' becomes superfluous) it has become politically important to conflate them in order to maintain a fragile consensus. Ironically, as the concept of sustainable development has become prominent, use of the word 'sustainable' has proliferated to describe something which can become self-maintaining over the long term without any reference to the environment or ecology. For example, in the UK, the Environmental Audit Commission accused central government of debasing the concept of sustainability by 'indiscriminately use[ing] it in formulations such as Sustainable Transport, Sustainable Communities and Sustainable Growth which are primarily socio-economic concepts' (EAC cited by Russel 2007).

Boulanger (2007) argues that SD is 'a very young discourse' which is still at the 'controversial stage' and that the differences in terminology, position and opinion are partially due to this. Donella Meadows, co-author of the influential and controversial book, *The Limits to Growth* (1972) also considers the difficulties of such terms and their meanings as part of a 'messy social transformation'. She implies that we will eventually develop terms, concepts and meanings as distinctions become increasingly important for effective communication (Dresner

2002: 66). This would help with common understanding of terms, but it would not necessarily mean a greater consensus on how to act, because people will always have different values and will continue to create new ways of articulating them. It is also important to note that our understanding of SD will change over time as the earth and society pose new challenges (Baker 2006: 7).

Therefore although a theoretical distinction between SD and sustainability is possible, it is often hard to separate them and their meaning in spoken and written communications (including this book) due to their conceptual overlap, different usages, and their volatility over time. The variety of framings for quality of life and wellbeing carry potential for extreme confusion and manipulation, even before we introduce the concept of sustainability, another highly contested notion.

Modelling the 'contested concept'

SD has quickly gained prominence as 'an essential reference, which concerns all public policies' (Rey-Valette *et al.* 2007). Due to its vast and ambitious scope, its status as a 'meta-narrative' means that it is a 'container' for a wide range of issues and differing opinions (Boulanger, 2007). Like the term wellbeing, this is both its weakness and its strength. That SD can be seen as an overarching societal value gives it enormous (potential) institutional power, but as such there will be many differing views of what it might be and how it might be achieved. Moreover, although fundamental disagreements exist at the more general theory level, at the lower levels of detail and application, SD becomes even more conflictual (Dresner 2002; Dryzek 1997; Buckingham-Hatfield and Evans 1996a). Many have argued that SD is a meaningless concept because in practice it is used by different parties to promote their own interests and this is one reason for its wide acceptance (O'Riordan 1988).

The literature recognises a crude division between an 'ecological modernisation' or 'reformist' approach to SD, where existing economic and social issues are re-focused to include environmental ones (weak sustainability); and a more radical view of SD demanding fundamental changes in political and economic systems and in human behaviour and attitudes (strong sustainability). Differences in interpretation of SD are often characterised along an axis of 'weak' to 'strong' notions of sustainability: weak SD accepts the substitutability of natural capital for other forms of capital; whereas strong SD rejects this and focuses on sustainability of natural resources (O'Neill 2011; Neumayer 2007).

The ecological modernisation conception of SD can be linked to the political modernisation movement in established western democracies. In this view, theories of SD are influenced by dependency upon markets to promote technological innovation to solve environmental issues, and also upon regulation to prevent environmental degradation damaging market processes (Blowers, 2003). This approach incorporates environmental issues into the normal production and consumption cycles, for example, recycling and the use of renewables. It relies on technological and managerial solutions based on the current free-market system, thus preserving the existing dominant discourse of economic growth.

Unsurprisingly, this approach is the dominant conception of SD in public policy. So far SD has largely succeeded in 're-framing the main "industrialist" discourse with more attention to environmental issues' (Boulanger 2007: 29). Therefore some reject the whole notion of SD, arguing that it is really a form of conservatism which will lead to restructuring but no real change:

> In practice sustainable planning continues standard practice, and thus offers the best explanation for its success. Sustainability is an ideology used to justify existing policy and social order.
>
> (Treanor 1996)

Those who support a strong view of SD criticise this 'business as usual' approach and argue for SD as a 'challenge to the established order' (Buckingham-Hatfield and Evans 1996b: 6). Keough (2005: 71) sees sustainability as a 'theoretical and moral wedge to challenge neo-liberal globalization'. For some of these critics, SD offers a radical alternative which would 'usher us in a new fundamental policy controversy about the very definition and conception of the common good and human wellbeing' (Boulanger 2007: 27).

These various approaches to sustainable development are fully explored elsewhere,[5] and there is no room to set them out in detail here: suffice to say that a wide variety of conceptions of SD exist. Many people have tried to map them. For instance Baker's 'ladder' of sustainable development approaches (2006) is one of many attempts to provide an overview of the range and nature of differing stances towards SD. While useful, it illustrates an inherent problem with such typologies; they tend to conflate the conceptually separate areas of public participation, equity, environmental protection and ecocentricity, and assume that if an individual upholds one of these then they automatically uphold the others. This assumption allows the author to order all the separate aspects of SD together along a single axis of weak to strong (Connolly 2007). Two-dimensional or 3D graphic models of SD are more useful because they allow a more nuanced range of views to be represented.

Various graphic models of SD have been produced as a tool for conceptualising this complex and wide-reaching theory. In the early days, SD was often described as a stool with three legs: environmental, social and economic, the three 'pillars' of SD. Given the inextricable relationships between these three domains the image is counter-intuitive. A more popular and enduring model of SD replaced it. The Venn diagram of three interlocking circles representing environment, economy and society (or sometimes equity) has become embedded, as Connelly (2007: 263) argues, as 'part of the taken-for-granted language of sustainable development'. Levett (1998) has criticised this 'three-ring circus' and the image it gives of each domain being separate and also equal in size and therefore by implication, importance. He prefers a 'Russian doll' model, arguing that the largest container is the environment which is in turn a precondition for society, which itself is a precondition for economy. Others argue for a fourth 'institutional' or 'political' sphere (Ayong Le Kama 2007; Rey-Valette *et al.* 2007). Although all graphics are ultimately defeated by the complexity of SD and therefore remain contested, the Venn diagram has endured and powerfully influences the way we talk about sustainability and conceptualise its measurement. It is important to understand that such models are not impartial technical representations of SD but are created by, and help to create, particular framings of it.

Measuring sustainable development

Like wellbeing and quality of life measurements, sustainability indicators are viewed as essential tools for understanding, promoting and operationalising SD:

> Until such time as there are measurement and assessment tools, no effective sustainable development policy can be set out.
>
> (Ayong Le Kama 2007)

Chapter 40 of Agenda 21 states that sustainability indicators are essential 'to provide solid bases for decision-making at all levels' (UNCED 1992). The development of indicators has subsequently become a worldwide industry and, as Boulanger (2007: 16) describes, 'the biggest game in town'. Before Brundtland, sustainability indicators (SIs) had been essentially based on natural science methods (Bell and Morse 2000). In the late 1970s and early 1980s, increasing concern about the environment created a situation where 'environmental indicators eclipsed interest in social indicators' (Carmichael *et al.* 2005: 178). These environmental indicators were essentially used for technical purposes such as mapping air and water quality, traffic flows, biodiversity and natural resources, and were seen as belonging to the domain of environmental technicians or experts. In the last 20 years, since the Brundtland and Agenda 21 reports, the scope of indicators has broadened to include social and economic conditions. Agenda 21's emphasis on local action, coupled with the central position of local people in sustainable development discourses, meant that participation came to be a pre-requisite in the development of community-based or local quality of life indicators. Local councils and municipalities embraced Agenda 21 enthusiastically, and many community quality of life indicators have been produced across the world[6] (Sirgy *et al.* 2004). The debate and learning that occurred in the process of developing indicators became as important as the product itself.

Local indicator projects reflect varying local views of what sustainable development/quality of life comprises, and measure anything from unemployment and smiling in the street to frog populations. Whether these projects describe their aim in terms of 'sustainability', 'quality of life' or 'healthy cities', the fact that they all employ economic, social and environmental indicators suggests they share similar goals. Yet despite this widespread activity (highlighted by the ongoing series, *Community Quality of Life Indicators – Best Cases*, see Sirgy *et al.* 2011), experts in both of the sustainability and wellbeing measurement fields scratched their heads. To researchers already deeply steeped in the difficulties of measuring either quality of life or environmental quality, the attempt to integrate another whole field seemed an almost insurmountable challenge. Environmental experts began to grapple with issues already explored by quality of life researchers:

> Pollution and erosion may be measured but how can quality of life be assessed? There are numerous examples of gauging wellbeing through employment, income, crime, travel, migration, house prices. However just which of these or others are important will presumably vary from individual to individual and over time. Calibration and interpretation would also appear to be problematic. Are they all to be treated equally or is crime to be rated higher than travel?
>
> (Bell and Morse 2000)

Conversely, in the field of wellbeing measurement, researchers started to tackle the challenge of including environmental quality as a key component in the concept and measurement of quality of life:

> At this point in time, it is clear that what is required is a system that accommodates not only economic and social indicators, but indicators of environmental degradation and resource conservation. In short, what is required is a comprehensive system of measuring the wide variety of aspects of human-wellbeing, as well as the means of improving it and sustaining it. Unfortunately such a system (as I imagine it, anyhow) would involve the construction of something like a general theory of a good society (something like a utopia) which would be generally acceptable to most people...This is practically impossible because we cannot get agreement on the elements of the utopia or on the proper evaluation of those elements.
>
> (Michalos 1997: 222)

Nevertheless, new sorts of indicator emerged with the aim of including SD in mainstream economic measurement and of setting an alternative to GNP. One leading example is the Index of Sustainable Economic Welfare (ISEW) or 'Genuine Progress Indicator' (Neumayer 2007). This index includes a large number of social and environmental adjustments and costs which are absent from GNP measurement. Innovative measurements of ecological footprints (Chambers *et al.* 2000) or environmental space (Opschoor and Wetering 1994) try to indicate in various ways the equitable allocation of resources within biophysical limits. New Economics Foundation, in conjunction with Friends of the Earth, produced the (un)Happy Planet Index (HPI) which assesses the economic and environmental efficiency of different nations to produce wellbeing by the equation HPI = (life satisfaction × life expectancy) ÷ ecological footprint (Marks *et al.* 2006a). While so far these newer measurements have not become institutionalised, some are gaining currency (Boulanger 2007). Although considerable problems with measurement remain, many local authorities are now considering the ecological footprint model, for example.

A little bit of everything doesn't make everything a little bit better

In 1997, the International Institute for Sustainable Development produced the Bellagio principles to review and support progress on developing indicators for sustainable development (Hardi and Zdan 1997: 2). The report recommended 10 core principles to guide indicator development at all levels, from community groups to international organisations. The report stressed the 'holistic' nature of SD, and said that indicators should:

> Consider the well-being of social, ecological, and economic sub-systems, their state as well as the direction and rate of change of that state, of their component parts, and the interaction between parts.

Despite the recommendation for measuring interaction of factors, many commentators are concerned that although most nation states and many communities have developed a set of indicators to measure sustainability, these tend to be 'lists of loosely connected economic, environmental and social indicators' with little reflection of trade-offs, and as a result these sorts of indicators can be seen as part of the weak 'business as usual' approach (Boulanger 2007). Rey-Valette *et al.* (2007) argue for indicators to be co-ordinated around key issues which link the aspects of SD, whereas currently they are predominantly based on the separate domains, resulting in an 'inventory logic' and do not aid the meaningful integration of differing views. As such the process of developing indicators tends to be driven by 'consultation' rather than 'debate' resulting in roughly equal numbers of indicators under each domain reflecting the assumption that this will result in sustainable development. However, scrutiny of the synergies and tensions between the separate but related domains of SD, and between the concepts of SD and wellbeing (however defined) are essential in the process of developing indicators.

Sustainable development, wellbeing and quality of life

Quality of life, wellbeing and sustainable development are complex and contested concepts and 'different for different people'. These are social and political constructs, and the harmony or conflict between them will depend on how they are understood, defined and manipulated in different contexts. For example, deep ecology literature considers human wellbeing and sustainability as one and the same thing, because humans are essentially ecological beings, physically, emotionally and spiritually dependent on the earth (Naess 1989). For wellbeing to increase, humans need to 'reconnect' themselves to the earth through 'self-realisation'. Apparently this creates relatively few tensions with the concept of sustainability. But even an ecologically-minded being has a body and needs that use resources. O'Neill (1993: 151) sees the view of 'holistic' wellbeing as deeply flawed because it tries to connect humans with a nature that is fundamentally unknowable and this 'strangeness' is what makes it so precious. He maintains that if we can develop a 'disinterested' appreciation of nature by 'humanising' our senses and developing skills of perception and observation of nature (rather than of ourselves) this will provide the key for an appreciation of its intrinsic value. To O'Neill, therefore, the ecocentric approach is deeply anthropocentric; there is no better way of creating a self-interested relationship with nature than by becoming one with it. He calls for the pursuit of science, art and crafts as the way to foster human creativity, excellence and practical reason and to take part in associations away from both the market and the state. He argues that this will both enhance human wellbeing and increase ecological awareness. His underlying assumptions about humanity are that when humans become truly or fully human, they will necessarily appreciate the environment. Nussbaum (2000) includes in her list 'being able to live with concern for and in relation to animals, plants, and the world of nature' as a central human capability. Gough (2003) views this inclusion in Nussbaum's list, and on an equal footing with 'basic' needs, as 'incredible', a critique which reveals different ontological beliefs. For a thorough review of the different ways the relationship between humans and nature has been perceived and analysed see Macnaghten and Urry (1998). Humans, of course, are ecological beings in a general sense but they also have social, cultural, political and individual needs and expressions. This makes factors such as status, property ownership, freedom of expression and autonomy important. That is not to say they couldn't change but at the moment many of them depend on or benefit from various forms of consumption, and so substantial 'trade-offs' are required for sustainability:

> To make sustainability happen, we need to balance the basic conflict between the two competing goals of ensuring a quality of life and living within the limits of nature. Humanity must resolve the tension between ultimate ends (a good life for everybody) and ultimate means (the capacity of the biosphere).
> (Chambers *et al.* 2000)

As already argued in Chapter 2, in western liberal democracies neoliberalism promotes the idea of quality of life, using the concept of the autonomous individual satisfying preferences through market-based mechanisms. As Nussbaum argues 'many critiques of liberalism are really critiques of economic utilitarianism' (Nussbaum 1999: 57, cited in Raco 2005: 328). However, our belief of what it is to be human is not just underpinned by neoliberal discourses but also by concepts of freedom, justice, equity and democracy. Wellbeing theory and measurement is predominantly located around the individual, their rights, capabilities, personal happiness and psychological fulfilment. It is almost impossible to conceive of a different way of constructing human wellbeing which would not only compromise deeply held normative beliefs, but also challenge our basic comprehension of what it means to be human. I argue that ultimately the concept of sustainability must pose serious problems for this construction of quality of life, because it asks us to change the way we conceive of the self. The need to live within the biophysical limits of the planet means that sustainability raises questions about our status as ecologically determined beings and must involve troublesome issues like population limits. In a liberal society, where the right to sexual and reproductive freedom is enthroned in basic notions of liberty and justice, these notions are dismissed as 'Malthusian' or 'eco-authoritarian'. Without going that far, it is easy to see how the new dimension of ecological limits poses problems for liberalism and why so much literature debates whether SD can in fact be reconciled with it. Some argue that because of the way autonomy is configured in liberal democracies sustainability cannot work without radical structural change, as SD looks more like a form of green socialism (Eckersley 2004; Dresner 2002). In any case sustainable development places 'unprecedented emphasis on the common good' and therefore 'heightens tensions with the private and personal' (Laessoe 2007). This tension between personal freedom or autonomy and the common good is a useful way to conceptualise the difference between wellbeing/ quality of life and sustainability. In much literature about wellbeing, particularly the recent SWB literature, insufficient attention has been paid to potential tensions between individual and societal wellbeing. Although a great deal of work has looked at the importance of social and community life to individual wellbeing, or social wellbeing, there has been little discussion on what societal wellbeing actually looks like or reference to social quality research. Sustainability requires us to think *as* a collective, rather than what we as individuals can get from or give to a community; it asks us to think about human wellbeing as a common project and to make sacrifices for people with whom we will never have a relationship. Even more difficult, the notion of ecological limits means that we should expect to have fewer individual rights and freedoms. If so, which rights or liberties must we forgo? Freedom in reproduction? In medical care? State support? Consumer choice? These things are not easy ones to consider. But neither is the fact that in Lesotho in 2011 average life expectancy was 48.

The differentiation of individual and societal wellbeing can be seen as reflecting the perennial struggle between libertarianism and egalitarianism (Phillips 2006). SD employs social equity, participatory and environmental justice as its key normative and instrumental principles, rather than individual freedoms and civil liberties. Despite strong discourses of democracy and participation it is rare to find an account of local sustainability which does not *also* mention 'education', 'facilitation' or 'management' of local people. For many the issue is predominantly about individuals having to change their views and behaviour. In the UK there is a reluctance to impose such changes through fiscal and legislative means. This is partly because of political risks but also because individual autonomy is considered one of the key factors of human wellbeing (Marks *et al.* 2006b). Thus liberal democracies which uphold individual freedom as paramount have an inherent governance problem. Behaviour change needs to be linked to autonomy, people have got to *want* to change (Zidanšek 2007). Foucault's theory of governmentality has been used by many scholars to explain how modern liberal governments manage populations through 'recasting subjectivities' to reproduce government agendas (Rose 1996). In other words, governments seek to create social 'truths' through discourse which people then internalise and act upon while still retaining a sense of autonomy. This concept has been applied to the way that environmental modernism is being promoted by governments producing forms of eco-governmentality or environmentality (Rydin 2007; Agrawal 2005; Baldwin 2003; Luke 1995).

The conflation of the terms sustainability, quality of life and wellbeing in much UK government rhetoric means that little critical attention has been paid (until recently) to the conceptual relationship between wellbeing and sustainable development. Despite the rising interest in both wellbeing and sustainable development in the UK, Marks *et al.* (2006b: 14) observed that 'in terms of public discourse, the wellbeing and sustainability debates have been held at some distance from each other'. The rest of this chapter focuses on the UK policy context.

UK government policy

In many ways, the UK has been at the forefront of strategy on SD just as it has with wellbeing. Following the Rio Earth Summit, the UK's Conservative government was among the first EU member states to publish an SD strategy, in 1994. In 1997, the new Labour government 'injected fresh impetus into the SD process and broadened its scope' to provide a more holistic view of SD, including economic and social, as well as environmental, criteria (Russel 2007: 190). SD enjoyed high-level ministerial support under Tony Blair's government. It produced its first national sustainability strategy in 1999, and a second in 2005 that used indicators developed in response to criticisms that the first showed insufficient understanding of SD and gave too much weight to the economic dimension at the expense of the environment. In 2005, 14 of the set were given the title 'headline' indicators 'to focus attention on what SD means'. The new indicators were generally accepted as 'painting a more comprehensive picture of the progress needed for SD' (Russel 2007).

In the early days, local authorities saw Local Agenda 21 as an innovative approach to the complex issues of SD, an approach that entailed collaboration and consensus between the local authority and local stakeholders. Reflecting the key SD tenet of participation, the involvement, education and empowerment of local people was stressed. Although evaluations of LA21 concluded that often the process was over-emphasised at the expense of concrete policy change (Buckingham-Hatfield and Evans 1996b), it was acknowledged that LA21 had effected a 'modest shift towards more participatory methods being adopted by municipal authorities' (Rydin 2003: 151). However, LA21 was perceived by the government as essentially a range of diverse, small-scale initiatives predominantly focusing on the environmental aspect of SD, and it sought to mainstream SD by setting out a more 'holistic' account of it to include the economic and social element (Russel 2007). Indeed, the government wanted to make it a 'cross-cutting' concern by embedding it into all strategic policy at a local level. A key instrument of this was the Local Government Act 2000 which was a crucial element in New Labour's modernisation of local government. Local authorities now had to work in partnerships with other agencies and with local communities to improve 'social, economic and environmental wellbeing' (DETR 2000). A local 'vision' of what this might mean for the area had to be articulated in a Community Strategy. The principles in forming this modernisation agenda were interlinked and interdependent with the government's desire to mainstream the principles of SD:

> Through its explicit agenda to 'modernise' local government, the Government is seeking to establish sustainable development as a core policy principle.
>
> (Evans *et al.* 2003)

Local quality of life indicators were seen as one mechanism to help local partnerships articulate their vision for the area, steer and inform policy and help evaluate progress towards SD. In the UK, local government modernisation and SD have both driven the development of QLIs (Shepherd 2007). It is important to understand that the notion of quality of life measurement was embedded in, and an instrument of, the UK government's agenda of social and political reform. There were considerable challenges in implementing this agenda. In order to explore the development and effectiveness of local indicators during this time it is important to set them within this political context. The next section sets out a brief account of this context and the relevant policy instruments.

New Labour's modernisation agenda

The Labour Government elected in 1997 sought to 'modernise' the public sector through a range of policies and interventions which were collectively called the local government modernisation agenda (LGMA) (Darlow *et al.* 2007). One of the new and central agendas for New Labour was that of 'joined-up government', and the LGMA had three central aims: modern services which are efficient and of high quality; democratic renewal; and community leadership where local governance creates a 'shared vision' for the area (Sullivan 2005). Ostensibly, this agenda was driven by what central government perceived as 'fundamental problems within the culture and framework of local government' including weak relationships between partners; services run according to council expediency rather than public need; indifference of local people; and seriously under-represented groups amongst councillors (Evans *et al.* 2003). The 1998 White Paper *Modern Local Government: In Touch with the People* made the issue clear: 'there is no future in the old model of councils trying to plan and run most services' (DETR 1998). The LGMA can be seen in the wider context as part of a global shift over the last 50 years in the way that governments work.

New Labour proposed radical changes to the public sector to invigorate local democracy. Its increasing emphasis on local communities as the sites not only of policy focus, but also of responsibility for policy, produced powerful discourses of community participation. The terms 'community' and 'participation' were identified as 'hurrah' words for New Labour, two of the 'positives in the pantheon of New Labour language for civil society' (Dinham 2006: 183). For New Labour, community partnerships were increasingly seen as the key to creating successful local policies.

Local Government Act 2000

The Local Government Act 2000 was significant as it imposed a *duty* on all local authorities to create local strategic partnerships (LSPs) within which to debate and decide local priorities. Geddes *et al.* (2007) have described this as a 'major innovation in the pattern of local governance' in that it:

- Brings together at a local level the different parts of the public sector as well as the private, business, community and voluntary sectors;
- Is a non-statutory, non-executive organisation;
- Operates at a local level which enables strategic decisions to be taken yet is close enough to the grassroots to allow direct community engagement.
- As the 'partnership of partnerships' in a locality, the ability of the LSP to provide an arena for community leadership and joined up service delivery is vital to the local government modernisation agenda.

(Geddes *et al.* 2007: 97)

The act also placed a duty on LSPs to produce a Community Strategy. This strategy is intended to be the 'plan of plans' (Darlow *et al.* 2007), something that must underpin all other strategies and to which they must refer. In March 2005 with the release of the new UK Sustainable Development Strategy, *Securing the Future*, community strategies were rebranded 'sustainable community strategies' which 'will evolve from community strategies to give a greater emphasis to sustainable development objectives' (Defra 2005).

Sustainable communities and local quality of life indicators

In 2003, John Prescott, then deputy prime minister, launched the 'Sustainable Communities: Building for the Future' plan, predominantly to address housing issues through a £38billion programme of regeneration. This was hailed as a major initiative to promote economic development within a sustainable framework and it had a strong focus on local community governance as a vehicle for guiding development (Raco 2005). This signalled the increased emphasis that New Labour put on sustainable communities as 'the essential building blocks of social harmony and progress' (Raco 2007: 308). 'Sustainable communities' became a core guiding principle for New Labour and informed the guidance for developing Sustainable Community Strategies in the UK. Although local strategic partnerships were encouraged to develop their own vision for the area, the Office of the Deputy Prime Minister (ODPM) now the Department for Communities and Local Government issued guidance about what a sustainable community should look like (see Box 3.2).

Community Strategies were intended to reflect this broad template of sustainable communities, and guidance for local authorities produced by Defra asserted that local problems should be identified and priorities decided through local quality of life indicators (Defra 2006). Intensive work was conducted at national and local levels on the (non-mandatory) development and promotion of

Box 3.2 Sustainable communities

Active, inclusive and safe – fair, tolerant and cohesive with a strong local culture and other shared community activities
Well run – with effective and inclusive participation, representation and leadership
Environmentally sensitive – providing places for people to live that are considerate of the environment
Well designed and built – featuring quality built and natural environment
Well connected – with good transport services and communication linking people to jobs, schools, health and other services
Thriving – with a flourishing and diverse local economy
Well served – with public, private, community and voluntary services that are appropriate to people's needs and accessible to all
Fair for everyone – including those in other communities, now and in the future

Source: DCLG website 'What is a Sustainable Community?'

local QLIs during the early 2000s. In 2000, DETR published a set of Local QLIs (DETR 2000). The Audit Commission (AC) was instrumental in taking this work forward and piloting and developing a set of local QLIs, working closely with the (then) Office of the Deputy Prime Minister (ODPM). They undertook extensive research[7] with local authorities to provide a set of indicators and guidelines for their use (Audit Commission 2002a; 2002b, 2002c; 2003; 2005). Their 2005 guidance on local QLIs lists 45 indicators and provides a database of evidence relating to each indicator for each authority. This list is included in Appendix A and was used as a reference for the case study.

QLIs were seen as effective tools for the Sustainable Community Strategies, and although the Audit Commission had developed a ready-made set, local authorities were encouraged to work with partners and communities to create their own at a local level. QLIs were required to fulfil various functions: to provide a vision for the area; to enhance local democracy; to provide information for effective policy formation; to evaluate those policies; and to increase education and action around sustainability issues. The process of creating new local partnerships and community strategies under this act meant that, in many places, Local Agenda 21 partnerships, strategies and local sustainability indicator sets developed in the 1990s were incorporated into or superseded by the new LSPs, Community Strategies and QLIs. The government intended that Local Agenda 21 programmes would be incorporated into the work of the LSPs but prescribed no specific mechanism by which this could be ensured. A study investigating how LSPs in London incorporated sustainability issues into their work stated the findings were 'not encouraging' (LGIU 2005). In the translation of LA 21 into community strategies many local authorities seemed to take a backward step in terms of sustainability. The study, carried out by the London Sustainability Exchange, found that quality of life indicators largely represented a local regeneration or 'liveability' index reflecting little local engagement with how local processes of development interacted with global concerns.

Where does wellbeing fit in?

In response to growing interest in subjective wellbeing, the Sustainable Development Research Network (sponsored by Defra) commissioned a discussion paper in December 2005 to look at the concept of wellbeing (McAllister 2005). This paper highlighted the need for research in the following area:

> To fully understand the relationship between wellbeing and sustainable development we need to examine the trade-offs between wellbeing now and later (intergenerational conflict); individual and social wellbeing (conflicts related to individualism and relative deprivation): the interaction between human, economic, and ecological wellbeing (how to achieve a sustainable balance).

Defra took up this recommendation and commissioned two pieces of research to specifically look at this relationship (Dolan *et al.* 2006b; Marks *et al.* 2006b). Dolan *et al.* (2006b) conducted a literature review and a study of policymakers' attitudes through a questionnaire and a workshop. They found a 'lack of common understanding about these conceptual differences' and argue, unsurprisingly, that synergies and tensions between the concepts of wellbeing and sustainable development will depend heavily on how those concepts are understood. Their report sets out a range of different understandings of wellbeing and compares various accounts to sustainable development approaches. Dolan *et al.* use 'institutional' SD (after Mebratu 1998) which aims for 'clean, equitable economic growth'. They compare discourses that include weak and strong sustainability accounts (which they term 'steady state' and 'utopian' respectively) and those emphasising environmental equity or intergenerational equity. The researchers split wellbeing into four main accounts:

1 Objective lists – based upon theoretical and intuitive accounts of wellbeing
2 Preference satisfaction – based on fulfilling desires
3 Flourishing – the satisfaction of psychological needs and reaching one's full potential
4 SWB – based on positive emotions and overall life satisfaction.

They use these different conceptions to explore the relationships between the various notions of wellbeing and notions of sustainability, and conclude:

1 Objective lists. Accounts of wellbeing based on objective lists appear consistent with SD if those lists include elements conducive to SD such as environmental protection. However, in such lists the trade-offs between the

different elements in the list are not normally considered and there is no clear consensus about what such lists should comprise. This limits their usefulness in applied policy settings.

2   Preference satisfaction. This seems to offer the least synergy with SD, because preference is expressed through the market, thereby increasing consumption. However, other preferences may exist which are not expressed through the market, and finding a way to measure these may reduce the tensions between preference satisfaction and SD.

3   Flourishing accounts. These provide no clarity because of ambiguity about what our psychological needs really are. For example, the focus on increasing synergy between wellbeing and SD through the reduction of materialism, does not take into account the need for social status and how that might then be manifested elsewhere.

4   SWB. Dolan *et al.* found synergies and tensions between SWB and SD, but felt that SWB was the most valuable of the four as a means to conceptualise wellbeing in public policy.

Their technical account is complemented by a report from the New Economics Foundation (Marks *et al.* 2006b). This strikes a more normative tone that reflects the foundation's strong sustainability stance and commitment to radical social and economic restructuring. It sees sustainable development as consisting of human wellbeing and ecological sustainability, and therefore considers the relationship between them, rather than that between wellbeing and SD. Marks *et al.* use a 'looser and more inclusive understanding of wellbeing' which combines a needs-based approach like Brundtland with a flourishing account, thus recognising both physical and psychological needs, chief among the latter being the need to feel competent, efficacious, free and autonomous, and the need for meaningful social connections with others.

Marks *et al.* present a typology of ways in which wellbeing (as defined) may affect, or be affected by, ecological sustainability, in direct, transparent and opaque pathways. For each pathway they set out a series of 'connects' and 'disconnects' between wellbeing and ecological sustainability. In Table 3.1, these are collated with the substantive synergies and tensions outlined by Dolan *et al.* (2006b).

A win-win situation?

What practical use can we make of the above attempts to create conceptual clarity for public policy? Dolan *et al.* concentrate on trade-offs between various aspects of wellbeing, and between wellbeing and SD. Marks *et al.* suggest that reconfiguring notions of wellbeing around a de-materialised economic system will deliver SD in a win-win situation. They use the argument that materialism and consumption will not only affect future generations but is already having an impact on wellbeing today. They claim that an emphasis on declining levels of wellbeing would have a powerful impact on policy, and therefore this should be the platform from which to argue for sustainability:

Table 3.1 Substantive synergies and tensions between wellbeing and sustainability (from Marks *et al.* 2006 and Dolan *et al.* 2006b).

Connects	Disconnects
Reduced car use, more exercise enhances physical and mental health, less air pollution, noise and accidents, more personal income	Car use supports freedom, autonomy and for some access to essential services
New jobs in renewable or eco industries, eco-tourism	Limits to economic growth in current system risk rise in unemployment, bad for wellbeing
Climate change and environmental degradation bad for wellbeing	Climate change for some limited areas means greater SWB
Preservation of green space and biodiversity important for wellbeing	Resource intensive behaviours support wellbeing in the home, promoting comfort and saving unpleasant tasks
Less fuel poverty, eco-friendly houses	Need to make trade-offs between health, income, freedom
Greater SWB correlates with environmental values	Inherent needs for social status means for many social status through materialism may offset greater SWB
Greater SWB correlates with post-materialist values	Comfort/luxury standards compromised
Greater SWB correlates with participation	Widespread availability of flights has increased wellbeing
Consumption growth has caused breakdown of social structures which support wellbeing	
Noise pollution from aircraft	

The potential for wellbeing for those promoting sustainability is clear. If unsustainable behaviours – indeed, if material consumption growth itself – were shown to be detrimental to wellbeing, in addition to their negative ecological impacts, this would represent an extremely powerful argument in favour of a move toward explicitly pro-environmental policies. Moreover, it would be an argument with genuine marketing appeal: selling sustainability, in these circumstances, ought to be a good deal easier than it has been to date.

(Marks *et al.* 2006b: 15)

This reflects ambitions to integrate the theories of human wellbeing (particularly SWB) and ecological sustainability. For example, using research by Diener and Oishi (2000) showing that happier people have less materialistic values (where of course constructs of 'happiness' and 'materialism' in such research are open to question), Zidanšek (2007: 896) makes the leap to claiming a strong correlation between happiness and sustainability. He also suggests that 'happiness could... play an important role in global efforts to reduce CO_2 emissions'. While acknowledging the complexity and problems of measuring both happiness and sustainability, he maintains that promoting individual sustainability values through education is the starting point for a transition from materialistic to post-materialistic values. A more pragmatic narrative by Gasper (2007b: 489) suggests that 'sustainability may be best promoted by appeal to people's self-interest: their eudaimonic and even hedonic prudence' rather than appeals on the grounds of intra- or intergenerational justice. However, Dresner is rightly cynical of such claims:

> As many Green authors have pointed out wealth and materialism do not bring true happiness. It is a truth that was pointed out much earlier by Buddha and then by Jesus. Yet over two millennia of this knowledge have failed to prevent consumerism from replacing both Christianity and Buddhism as the religion of both Europe and East Asia.
>
> (Dresner 2002: 171)

Dolan *et al.* and Marks *et al.* concentrate heavily on influencing individual behaviour. Autonomy is discussed by both reports as being intrinsically important to wellbeing and this poses a wellbeing challenge to the imposition of financial and legislative penalties on anti-environmental behaviour. Dolan *et al.* argue from an instrumental point of view that 'shifting attention to improving social relatedness' creates more opportunity for social norms to influence pro-environmental behaviour (as outlined by Jackson 2005) and has the added advantage of 'focussing attention on the common good rather than individual benefits'. Marks *et al.* argue from a slightly more normatively-infused perspective that social relatedness is one of the fundamental constituents of human wellbeing, and that it has been eroded by increased individualisation in society, leading to the weakening of family and community cohesion and increasing vulnerability to depression and isolation. Their study sees this as the key new wellbeing message that will drive a shift into more ecologically sustainable policy and behaviour through an increasing dissatisfaction with the current model of economic growth. Marks *et al.* (2006b) see a focus on human wellbeing as a way to move beyond the pro-growth/anti-growth debate where economic growth is seen as a means rather than an end in itself. They place a strong emphasis on wider structural limitations to sustainability and the need to de-couple economic growth from production and consumption. This argument has been taken further more recently by Jackson (2009) who argues that decoupling has not worked fast enough and that economic growth must instead be more closely related to those activities which enhance public and environmental goods. But he is not explicit on the relationship between public goods and environmental goods, or the potential trade-offs between them.

The ideas of a win-win situation between sustainability and wellbeing through social cohesion and social capital, gained currency with central government and were seen as some of the key tenets of New Labour. The idea of 'community' was seen as the answer not only to individual happiness and sustainable development, but also to local democratic renewal and devolution of power. In his work on pro-sustainable behaviours, Jackson (2005) views communities as one of the main avenues of creating new forms of behaviour because of how they can exercise influence upon members. He argues that small groups or communities are the context within which the most powerful social norms are applied. One important factor to small-group management is participatory decision making; if people have participated in the creation of group norms, they will more effectively internalise them and these norms become a part of the group's shared meaning. Jackson recognises the cultural basis of consumption and the power of the private sector, so he focuses on peer groups and communities as the basis for change (Jackson 2006). The link to sustainable development, however, relies on which norms are promoted and internalised. Jackson (2005: 126) sees a potentially positive role for the state 'as a continual mediator and "co-creator" of the social and institutional context', in turn referring back to top-down discourses of sustainable development. This is a pragmatic narrative, but while it acknowledges current structural limitations, its focus on change at the local level both reflected and fed into New Labour discourses about 'community' and 'participation'. This was also reflected in the language of NEF:

> As well as feeling satisfied and happy, wellbeing means developing as a person, being fulfilled and making a contribution to the community.
>
> (Shah and Marks 2004: 2)

'Making a contribution to the community' (depending on how defined) has been shown in several studies to be linked with SWB but causality is not clear. Do people volunteer because they are happy or are they happy because they volunteer? Marriage and religious faith are also positively correlated with SWB but NEF do not offer prescriptions for these, presumably as this would infringe civil liberties and human rights. But the expectations that we should contribute to our communities, and that this will make us happier, are assumed to be unproblematic. It is important to note exactly where messages about wellbeing are coming from, and how they are being configured to include notions of 'community', and how community is linked to SD. The explicit claim for 'contribution to the community' (rather than for social values in general) dovetailed neatly with New Labour's constructions of 'community' and for particular expectations and forms of behaviour. This will be discussed in the next chapter.

Summary

In summary, SD has emerged as a 'dominant discourse', the terms of which have been cast differently by different actors with different interests (Dryzek 1997). Most nation states have signed up to the Brundtland vision of the 'simultaneous and mutually reinforcing pursuit of economic growth, environmental improvement, population stabilization, peace and global equity, all of which could be maintained in the long term' (Dryzek 1997: 126). Within this win-win definition of SD, which is almost universally endorsed as a good thing, there is a wide variety of differing interpretations. Quality of life and wellbeing are underpinned by notions of economic growth. They are also deeply embedded in liberal societies which seek to promote the freedom and autonomy of citizens, and liberalism upholds the primacy of the individual's privacy and ability to act according to their own ideas of the good life. This causes problems for (strong) sustainability. However, a new discourse of SD-compatible or sustainable wellbeing (informed by SWB evidence) is emerging, informed by the scientific discourses of subjective wellbeing. This post-materialist discourse stresses 'goodness' or 'meaning' of living a certain kind of life and prescribes volunteerism and community participation as good for us. However, this is essentially a conflicted discourse; we engage in the community, not because we should (as that message would conflict with our freedoms) but because it will make us, as individuals, feel better. These two liberalist and communitarian discourses existed side by side in New Labour, the one upholding liberal economics and pursuing individualism and freedom, the other emphasising 'community' and responsibility, both seeking to co-opt the legitimising concept of SD. These two discourses are apparent in the Audit Commission list of Local quality of life indicators (see Appendix A) and lists of indicators driven by both notions were constructed by local authorities with no critical analysis of what was being measured or how they related to each other. The evidence indicates that the SWB model fits better with certain aspects of SD. However, as argued in the last chapter, SWB is not a wellbeing theory but a collection of complex and sometimes contradictory evidence, which like most evidence, is used selectively within specific discourses.

4 Leave it to the people?

Power and participation

Politics, like other arts and sciences, must consider not only the ideal, but also the various problems of the actual.

(Aristotle *Politics* IV.1)[1]

The previous chapters explored the various conceptualisations of wellbeing, quality of life and sustainability. These are all contested concepts, mediated through discourse, and therefore, fundamentally linked to democratic debate and an analysis of power. Aristotle thought that democracy was a corrupt form of constitution where a majority rule in their own interests as opposed to the interests of all, thereby ignoring the needs of a minority. A recent UK government strategy report noted that there is a 'widening gap in political interest between different social groups' and went on to say that 'particular attention may be needed in future to engage disadvantaged groups in the issues that affect their lives and involve them actively in the democratic process' (UK Cabinet Office 2008: 169). If disadvantaged groups are increasingly disengaging from politics, and inequalities are widening, how are their interests to be represented? Aristotle puts forward an interesting complex of two questions, still relevant for us today: whose interests are promoted in what sort of democracy?

Many political science authors point to a 'democratic deficit' in western liberal systems where political participation is flagging (Zittel and Fuchs 2007; Dalton 2004; Miller *et al.* 2000). Although there is some dispute about this in the UK, where studies have shown there are still considerable levels of political activity (depending on how this is defined and measured), its forms have become less traditional (e.g. election turnout) and more protest-based (e.g. joining a demonstration) (Carman 2010; Pattie *et al.* 2003). The literature on social capital reflects similar arguments about the possible decline in civic associative behaviour (Putnam 2000; Hall 1999; Maloney *et al.* 2000). Although the incidence of civic action is disputed it seems clear that the type of behaviour has become more 'commodified' (Grenier and Wright 2006). Also clear is that alienation from and distrust of political institutions is a key element in these trends (Carman 2010). Larger shifts over the last half-century towards more participatory forms of democracy in the UK are related to the changing relationship between the state and the citizen, and to expectations about the role of local authorities and their residents. Governments have tried to reinvigorate political participation through various means including more participatory methods. Yet, in the UK recent changes in governance and the emergence of New Labour's Third Way and the Coalition government's Big Society discourses around increased localism, community participation and social capital have been particularly criticised for making local people responsible for problems caused by extra-local systems and processes (for example Amin 2005). In this light, the current chapter considers the problematic concept of participation in local governance and the implications for defining and measuring wellbeing through local indicator development. I argue for a greater focus on the detailed processes of power than has hitherto been present in analyses of indicator processes and effectiveness, a greater reflexivity on the part of policy actors and researchers, and for greater value to be accorded to the discursive role of indicators.

I am convinced that in the attempt to develop local indicators, process is as important as product. Furthermore the quality of that process is important. Outcomes are more generally accepted if the process is perceived as fair, even if the outcome was not the one desired (Tyler 2000). It is only through a process of debate and discussion that we can forge a shared concept, not only of what wellbeing is and how to measure it, but also of what is fair and reasonable in deciding what and whose values should prevail. Moreover, a common claim for participatory processes is that resulting indicators will be more effective because they are 'owned' by a wider group of stakeholders and as such have greater social meaning, increase political trust and are linked to action (Sirgy *et al.* 2004; Lingayah and Sommer 2001). Collaborative processes therefore have potential to enhance local democracy and increase local capacity to address the complexities involved in assessing wellbeing and its relationship to sustainable development. Such capacity is generally low in local government, which is one of the main reasons cited for lack of progress towards sustainability (Evans *et al.* 2003).

However, participation is in itself a contested concept, subject to a wide range of understandings and discourses. For example, on the one hand it is promoted in new participatory planning paradigms (see Healey 2006 for example) where it is invested with a 'transformative power to effect social change'; on the other it may be a hand-maiden to central discourses, particularly as 'the rhetoric of participation dovetails nicely with the neoliberal agenda's emphasis on local control and autonomy' (Cahill 2007: 2861). Critics of local participation practices claim that the predominant idea of participation reflects an 'obsession with the local', a 'poor understanding of how power operates', is indiscriminately used on the poor and there is a tendency to treat it as a technical method rather than a political project (Terry 2008: 224; Cooke and Kothari 2001). This chapter will discuss these normative and critical claims, and acknowledge that 'we must "locate" participatory theory and practice, not as an ideal but as a messy, slow, engagement of power, politics, and context' (Cahill 2007: 2861).

I will therefore consider the problematic concept of participation in local government policy and practice with particular emphasis on the development of local wellbeing measurement through community indicators. First, I briefly outline the definitions and typologies of participation and its normative and instrumental benefits. I then consider it within the political context of the shifts in local governance and the emergence of New Labour sustainable communities discourses during the 2000s.

Definitions of participation

Participation, in the context of this book, is defined broadly as the involvement of citizens[2] in local policymaking. It suggests a mutually beneficial arrangement between policymakers and citizens which facilitates the influence of local people on decisions. Participation is important on both normative and instrumental grounds. Inappropriate policies which are foisted on local people pose a problem, not only for social justice but also in terms of efficiency and effectiveness:

> The justification for greater public participation is twofold: social justice on the one hand and functional legitimation on the other. If people feel that they 'own' the decisions made, then they are more likely to want to comply with them.
>
> (Buckingham-Hatfield and Evans 1996b: 10)

Participation is often described as a way to 'empower' people who are 'marginalised or disempowered' by producing new forms of knowledge for policy which would not otherwise be accessible and which can bring about social change (Jupp 2007: 2832). For some, like capability theorists, participation is a key part of their account of wellbeing, not simply a means to achieve consensus. Martha Nussbaum's set of central human capabilities includes, 'Being able to participate effectively in political choices that govern one's life' as one of the key criteria for a minimum account of quality of life based on social justice (Nussbaum 2000 and see Appendix B). Instrumental benefits of participation can include the development of more locally appropriate policies, increased stewardship of projects by local people, conflict limited by early involvement, a positive image and political legitimacy for policies and the creation of greater trust in government, that in turn facilitates shifts to participatory democracy (Adamson and Bromiley 2008; Kjaer 2004; Wagle 2000; Warburton 1998a).

Participation and sustainability

Participation is embedded in SD international protocols and is a central idea in the international sustainable development agenda produced by the 1992 Earth Summit at Rio.[3] Out of the 27 SD principles, four specifically relate to the benefits of 'local people' being involved in decision-making, including particular provision for the young, women and indigenous groups. For example, Principle 10 states that 'Environmental issues are best handled with the participation of all concerned citizens, at the relevant level' (UNCED 1992). This was further strengthened on a European level by the Aarhus Convention in 1998 which stated that the public had a right to participate in environmental decisions that affected them (UNECE 1998).

As set out previously, the fact that SD is a contested concept is often seen as a problem for community involvement and indicator development. However, if SD is viewed as a political construct with public debate at its core, its ambiguities become its strength:

> [The] lack of definitional clarity and unanimity of purpose should not discredit sustainability as a political goal and policy objective – on the contrary, the fostering of a lively and informed debate is likely ultimately to benefit the move to a more sustainable world – since we maintain that sustainability is, at its very heart, a political rather than a technical construct.
>
> (Buckingham-Hatfield and Evans 1996a: 3)

This positive, agonistic approach could facilitate a shift towards debate where these ambiguities could become a 'creative resource in the political process' (Myerson and Rydin 1996b: 33). A healthy local democracy is not only vital to the development of SD; the debate about SD can be a sustainable activity in itself. The success of SD in driving policy fundamentally relies on the participation of the local people. Local Agenda 21 was influential in helping to bring participatory processes into mainstream national and international policy (Warburton 1998a). Participation is now widely regarded as important for social justice and better policymaking.

Participation and subjective wellbeing

There is evidence that democratic participation is positively associated with SWB. Scholars claim that the acquisition of new skills, knowledge and confidence lead to greater satisfaction (for a review of evidence see Dolan *et al.* 2008). This association is strongest with forms of participatory democracy, for instance, in Swiss cantons where referenda are more common (Frey and Stutzer 2002; Stutzer and Frey 2006). Empirical research by Dorn *et al.* (2007) suggested that higher levels of democracy increase happiness. They argue firstly, that a greater level of democracy means that policies are more likely to represent people's wishes and therefore enhance their wellbeing; and secondly, that perceptions of procedural fairness may produce higher levels of SWB than the actual political outcome. Conversely, negative experiences of participation have an adverse effect on SWB levels. So although causality is not clear (whether participation makes people happier or happy people participate more) there seems to be a positive correlation between participatory democracy and SWB. Although the correlation between economic growth and increased levels of SWB is more tenuous, SWB does tend to be higher in richer countries and some have attributed this apparent contradiction to the comparatively high levels of freedom, human rights and democracy in such countries rather than their income levels (Dolan *et al.* 2008).

The problems with participation

Participation is not only important in the process of developing robust, locally resonant indicators for wellbeing or SD, it should also be included as an indicator because it may enhance both wellbeing and sustainable development. But despite the strong normative and practical reasons for promoting participation, empirical research consistently shows there is a gap between rhetoric and reality and expresses cynicism about much participatory practice (Jupp 2007; Cooke and Kothari 2001; Dinham 2006; Wright *et al.* 2006; Taylor 2007). Studies find that despite international discourses proclaiming participatory democracy, the reality is that local people are often ineffectively involved. This can lead to disillusionment and charges of 'tokenism'. There is often no explicit link between information gained from public consultation and decision making. This leads to processes that lack transparency, and to questions about how policies are made. The focus of these critiques varies as do the solutions offered: some recommend refinement of participatory tools, methods and learning; some scrutinise local institutional norms; some examine the wider structural and power issues at national or international levels. Some of the common issues concerning local governance are briefly outlined below.

First, creating structures for community participation does not necessarily ensure it, and often merely replicates established structures of representative governance (such as election) which discourages participation. Individuals vary in their capacity and preferences for particular types of engagement. Participation should fit within everyday informal networks such as book clubs, toddlers groups, gyms, faith groups etc. (Bush 2005; Wright *et al.* 2006; Adamson and Bromiley 2008; Skidmore *et al.* 2006). The argument, and it is disputed by some, is that political participation should be embedded into the social fabric of our lives. Second, the skills, knowledge and language of participants determine their ability to influence. Government literature often highlights the acquisition of skills as a positive feature of participation, as it helps the development of capabilities useful in the job market. People may initially need support, for instance, in organising a public meeting or a residents' group, writing a constitution, or understanding planning processes. The local authority often supplies this through community development work and can be a real advocate and facilitator for local voices; but in its role as 'gatekeeper', it may also make decisions that lack transparency about who is and is not suitable to speak. Local community participation can be dominated by a small group of 'usual suspects' who are attractive to formal governance because they have the prerequisite skills, knowledge and language but who may not adequately represent the range of local interests. They may also be disproportionately involved in activities, which can preclude others from the arena. These people may feel burdened ('If I don't do it, no one will') and this can lead to burn-out (Voisey *et al.* 2001; Adamson and Bromiley 2008; Skidmore *et al.* 2006; Wright *et al.* 2006). Third, participation in politics, as in other civic activities, closely correlates to social class (Grenier and Wright 2006). Last, problems in maintaining levels of participation are explained in terms of 'apathy', 'infighting' and 'lack of progress', where problems in collective action are often perceived to require top-down intervention in projects (Wright *et al.* 2006: 347). Projects can be long and complicated and many participants' expectations of what they can achieve diminish over time; this brings a risk of frustration leading to lower levels of participation (Dinham 2006).

Typologies of participation and power

In order to address some of these concerns, participation typologies attempt to highlight what 'true' participation looks like, and to separate the rhetoric from the reality. In Arnstein's (1969) widely quoted paper she identified eight categories of citizen involvement on a 'ladder of participation' and her contribution was to identify the rungs which were 'non-participation' or 'tokenism' (five of the categories) on the premise that citizens had no power to change anything. She classes practices such as 'information giving' and 'consultation' as tokenism because local people have no input into decision-making and no guarantee that their views will be heeded. She argues that surveys and public consultation are extractive rather than collaborative practices, with little community control of how the collected views are used. At the highest rung of participation on her ladder is 'community control', and she defines citizen participation simply as 'citizen power'. Arnstein describes a simplistic division between 'have-not citizens' and 'powerholders' but justifies this by arguing this is how these parties see each other, as a homogenous group and a monolithic entity respectively. While recognising underlying complexities obscured by her account, she bases her typology of participation firmly around a polarised view of power divisions. Arnstein was writing at the peak of an international civil rights movement when the concept of power was employed in discourses promoting radical change and encouraging grassroots initiatives. Inherent in these discourses was a commentary on social justice which went wider than the local community and invoked change to socio-political structures and processes. In the latter years of the 1960s 'the word participation became part of the popular political vocabulary' (Pateman 1970: 1). Empowerment of local people and social change in favour of the 'have-nots' was participation's holy grail.

Since then participation has moved from the margins to the centre of government policies and is gaining institutional legitimacy (Pain and Kindon 2007). This produced a new emphasis on the partnership between local authorities and 'empowered' communities as the ideal outcome of participation, but little real scrutiny of local power relations (Fremeaux 2005). As participatory practice has become a central part of government narratives, the discourse of power transfer from haves to have-nots has been modified to become simply the 'empowerment' of communities, often without detailed analysis of where the power will come from or how it will be relinquished by the 'haves'. Fairclough (2000:162) describes this process as 'nominalisation' where a social process is represented as a noun (rather than a verb) thus potentially side stepping any discussion of agency or responsibility; this was a key feature of New Labour language.

Davidson's (1998) reworking of Arnstein's ladder into a 'wheel of empowerment' for South Lanarkshire Council sought to avoid hierarchies of involvement based on a model of power-holding, by using a model where the 'right' options could be chosen to achieve 'identified objectives'. It still stresses the desirability of practices that are empowering, but has little of the normative force of Arnstein's proposal, taking a much more practical approach and focusing on 'transparency' and 'appropriateness'.

> The wheel promotes the appropriate level of community involvement to achieve clear objectives, without suggesting that the aim is always to climb to the top of the ladder.
>
> (Davidson 1998: 15)

It is now widely accepted that empowerment of communities is an often unfulfilled ideal and that it is impossible to include everyone at a truly participatory level. A guide produced by the New Economics Foundation for working with communities to produce a set of sustainability indicators adopts a pragmatic tone typical of such guides:

> Securing effective participation in local decision-making is an art, not a science; and a difficult art at that. It is 'politically correct' to claim that a project is using only the most participative and supportive forms of involvement.... There are many reasons why it may not be possible to be as inclusive as you would like; the real danger is *pretending to be more participative than you actually can be.*
>
> (original emphasis; MacGillivray *et al.* 1998: 21)

A similar view is expressed in a report evaluating a community indicator project in London where participative forms of involvement are described as 'not always practical or desirable' and which states that the 'need to be honest' is 'most important' (Lingayah and Sommer 2001: 13). The overriding principle promoted in these guides tends not to be empowerment but transparency. The recommendations are seen as realistic accounts of what is possible usually due to limitations of time, resources, skills and motivation of participants and facilitators. They do not necessarily engage with the wider structures within which participation (or the lack of it) occurs. Indeed, it is rare to see a detailed engagement with power in accounts of local indicator development, because participation tends to be treated as a method or technology at the level of community, rather than a political idea or part of societal culture.

Power

When the issue of power is raised, accounts tend to imply it is an entity which can be held and transferred (for example 'power holders' and 'power sharing'). However, a deeper engagement with the notion of power would result in a much more complex (and contested) construct, as the range of academic efforts to study it show (for example Lukes' 2005 classic work or Haugaard 2002). For instance, the elected leaders of a council argue for w. Council executive officers argue for x. A majority of the local public wants y. A minority of the public wants z. Here the councillors may exercise their democratic mandate to lead on decisions. The council officers may exercise power by appealing to rational argument and scientific evidence. The majority public may exercise power by exploiting the councillors' desire to stay on the right side of their voters or more directly through their democratic rights to vote them in or out. The minority public may exercise power by claiming they are being discriminated against and co-opting support from voluntary agencies who advocate powerfully on their behalf. It is easy to see that within this scenario different sorts of power may work in different ways over different timescales to produce decision (or non-decision). In addition, what may be more important than who holds power (since as we can see from this example power is not a defined 'thing' but something much more slippery) is how power is exercised and what the effect of that power is or might be within the local area. It is common in local government for policymakers to claim democratic legitimacy for their decisions by arguing that they have consulted with the 'community' or to claim scientific legitimacy by appealing to evidence. It may be more useful, then, to consider power located within discourse rather than within individuals as it is in appeal to 'truth' created by dominant discourses that people can claim legitimacy for decisions (Foucault 1991b). Yet, as previously argued, different discourses create different truths and are legitimate in different arenas and therefore the idea of power as an exercise working in complex ways through networks in society is appealing. Power is also productive as well as destructive and so the question is not only who exercises power but why and what impact it has.

The state/citizen relationship and the institutional context of participation

The rolling back of the state and the rise of diffuse partnerships has caused an 'institutional gap' (Hajer 2003) but institutions become even more important to maintaining a stable civil society in the face of increasing economic, social and environmental change, uncertainty and challenge (Jackson 2009). The perceived relationship between citizens and the state in the UK has shifted in the last half-decade. This has multiple implications for the role of local government. Historical and cultural analysis can tell us a great deal about how participatory democracy discourses operate in different contexts. In a local authority context, new discourses of participation overlap with understandings of governance, including entrenched hierarchies and traditional ways of working (Barnes *et al.* 2004). The following sections will explore how participatory practices have emerged in the UK, and how expectations have changed about the relationship that local authorities have with communities and citizens. While local institutions may be unique entities in themselves, they are also strongly influenced by national agendas. The case study in this book occurred during the New Labour Government so the following discussion will focus on that period, but within a wider historical context.

From clients to customers to communities

For most of the twentieth century the public were viewed as a collective of electors and ratepayers and the local authority/public relationship was a professional/ client one (Stewart 2000). Largely, the public were not seen to have a direct role to play in policy formation. The professionalism of local authority officers was the key mechanism for formation and delivery of policies. These professionals saw themselves as 'being at the forefront of a transforming effort, building the welfare states which would deliver a reasonable quality of life to the majority of citizens' (Healey 2006: 8). This professionalism was increasingly challenged during the late 1960s and 70s when unemployment figures rose to pre-war levels and post-war regeneration failed to deliver the expected social benefits. The public became more sceptical of local government, and political scientists described a 'participation explosion', partly as a reaction to this growing cynicism about the 'reign of expertise' (Cook and Morgan 1971: 1). Although these participation discourses gained popularity they made little impact on classical political theory and some warned of the dangers of participation to the stability of the political system (Pateman 1970).

At this time participatory approaches were developed primarily at the edges of government, in the community development field in order to work with issues such as poverty and poor living conditions (Warburton 1998a; Wright *et al.* 2006). They were strongly linked to grassroots action through anti-poverty campaigns and notions of the rights of disadvantaged communities (Fremeaux 2005; Pateman 1970). Participation became a formal criterion of many government grant schemes as it gradually gained legitimacy, although this legitimacy was earned partly because it was seen as a way of 'subduing potentially troublesome elements' (Hoggett 1997: 9 cited in Fremeaux 2005: 267).

A new, more corporate style of local government which emerged as management theories were implemented in the 1970s (Darlow *et al.* 2007; Stewart 2000). This was not just a UK phenomenon; commentators have observed a change in government style throughout the world at this time, in response to new patterns of work caused by globalisation. They have termed this the New Public Management (NPM) model. Although some authors refute the claim that NPM is a coherent model on the basis that its manifestations differ between countries, all these diverse approaches reject 'traditional bureaucracy' (Stoker 1999: 3) and emphasise 'efficiency, economy and effectiveness' (Astleithner *et al.* 2004: 11).

The Conservative Government in the UK (1979–97) pursued these ideas vigorously, introducing a system of market principles whereby members of the public were increasingly regarded as customers (Stewart 2000; Darlow *et al.* 2007). Compulsory (some might say compulsive) competitive tendering (CCT) was introduced as the mechanism by which local council services would be contracted out to the most economically efficient tender, and the government worked hard to redefine the relationship between the public sector and citizens. This government was keen to reduce citizen dependency on the state and encourage individuals to take more responsibility for their own welfare (Darlow *et al.* 2007).

In the 1980s, radically different discourses on participation appeared simultaneously. Right-wing ideas about active citizens as independent from the state and as customers of increasingly contracted-out services clashed with left-wing local authorities that supported the empowerment of local groups around issues like race and gender (Fremeaux 2005). The vigorous promotion of an 'enterprise culture' caused deep rifts in society at that time between individualist and communitarian discourses (Thompson 1992). Participation was linked with communitarian, left-wing politics and was seen by some as a challenge to representative democracy whose citizens agree to trade direct involvement for elective representation (Boaden *et al.* 1982). For these reasons participatory approaches had long stayed at the margins of policymaking. During the late 1980s and 1990s the failure of market-based initiatives to deal with inner-city problems, and a change in leadership of the Conservative party from Margaret Thatcher to John Major signalled a move from the aggressive promotion of enterprise culture towards 'softer' community and partnership discourses. New policies, such as the City Challenge and Single Regeneration Budget, began to reinstate the central role of local authorities in regeneration initiatives and to strengthen the role of community participation (Marinetto 2003).

When New Labour came to power in 1997 community participation was a core principle of their local authority modernisation agenda. They replaced CCT with Best Value, which was radically different in that it sought to involve local people in evaluating services and was designed to promote joined-up government between local authority, private and civil society (Marinetto 2003; Voisey *et al.* 2001). New Labour also made clear commitments to 'experiments in democracy' and employed deliberative mechanisms like focus groups and citizens' juries to engage citizens in policy formation and appraisal (Fairclough 2000; Taylor 2007). In the White Paper *Strong and Prosperous Communities* (2006) local authorities were seen as the 'nurturer' of local deliberative democracy (Barnett 2011: 278). New Labour used the term 'double devolution' throughout 2006 to indicate their wish to transfer responsibility for decision-making from central to local government and then from local government to citizens and communities (Adamson and Bromiley 2008; Taylor 2007). Local authorities were required to create local strategic partnerships and to demonstrate community involvement in the creation of (Sustainable) Community Strategies and Local Area Agreements, which were assessed in terms of 'community empowerment outcomes' (Taylor 2007: 299). Quality of life indicators were promoted as a tool to help deliver a shared local vision for an area, that reflected residents' views and values.

In a speech to the Labour Party Conference in October 2001, Tony Blair stated that 'the governing idea of modern social democracy is community'.[4] The concept of 'community' was a central notion in the ideology and policies of New Labour (Marinetto 2003; Fremeaux 2005). Blair and Brown were both heavily influenced by the communitarian ideas of Etzioni and the Third Way approach of Anthony Giddens (Fremeaux 2005; Terry 2008).

The idea of community was seen as a way to temper the power of both market and state in 'Third Way' politics (Giddens 1998). This ideology, promoted by New Labour, calls for a new partnership between the state and civil society with the aim of rejuvenating social cohesion and 'acting as a countervailing force to the excessive individualism fostered by modern global capitalism' (Marinetto 2003). In Third Way politics there was a strong expectation that individuals would identify with this notion of 'community' and become actively involved in their community (Etzioni 2001). Thus community empowerment is strongly associated with the idea of active citizenship and is also seen as one of the solutions to an increasingly squeezed welfare state:

> Cultivating communities where they exist, and helping them to form where they have been lost, is essential for future provision of much social good: it should be a major priority for future progress along the Third Way. In the next few years communities should be increasingly relied on to shoulder a greater share of those social missions that must be undertaken, because – to reiterate – communities can do it at a much lower public cost and with greater humanity than either the state or the market.
>
> (Etzioni 2001: 10)

Robert Putnam's work on social capital fitted neatly with New Labour's Third Way discourses. The publication of his book *Bowling Alone* in 2000 made him 'an almost-instant intellectual poster boy' for New Labour (Browne 2008). Social capital became a major policy aim across many departments and the Office of National Statistics created a unit specifically to measure it (Grenier and Wright 2006). Building social capital, particularly within disadvantaged communities, was widely seen as a way to increase access to resources through networks for those who lack 'the formal economic power to buy their way out of problems' (Skidmore *et al.* 2006).

> Social capital is attractive to policy makers because it holds out the possibility of improving social outcomes more effectively, through means that are more legitimate and cheaper than traditional public service alone.
>
> (Skidmore *et al.* 2006)

These discourses of active citizenship, community and participation tended to focus particularly on disadvantaged communities and were regarded as key in creating solutions to local problems (Taylor 2007; Raco 2005; Amin 2005; Jupp 2007). The 2001 National Strategy for Neighbourhood Renewal put 'community engagement at the centre of programmes to close the gap between the most disadvantaged and the rest of the country' (Taylor 2007: 299). The government established New Deal for Communities (NDC) in 39 of the most disadvantaged areas where residents were expected to play a key role in the delivery of projects.

Government literature typically portrayed community participation as instrumental to solving individual disadvantage as well as to promoting community cohesion:

> As well as contributing to improving the quality of life of their community, getting involved in local affairs can provide individuals with opportunities to acquire training, skills, and give them pathways into education and employment.
>
> (ODPM 2006)

But despite these very clear messages, empirical research records significant disparity between the rhetoric and the reality.

> There is little experience of the policy changes and the shifts of practice required by local and central government to deliver greater community empowerment in the UK.
>
> (Adamson and Bromiley 2008: vii)

The championing of participation became part of the politics of performance. Participation was compromised by the need to deliver actions and meet targets set by central government, and New Labour was slow to transfer real responsibility and funding to the local level (Dinham 2006; Wright *et al.* 2006; Marinetto 2003).[5]

Discourses of participation

The above description of these relationships between central and local government and the public is necessarily brief and general. The risk is that such accounts mask the actual complexity and diversity of relationships (Stewart 2000). Shifts in governance have produced increasingly hybrid attitudes to, and relationships with, the public that are hard to categorise. These shifts have been described as 'a reconfiguration of relationships and responsibilities, encompassing complex alliances of actors and networks across permeable institutional boundaries and an expanded vision of the public domain' (Taylor 2007: 299 citing Cornwall 2004: 1). They are also associated with normative concepts of 'community, social capital and civil society as integrating forces built on networks and trust' (Taylor 2007: 300; Kjaer 2004). Research conducted at the turn of the century to explore the effect of new forms of governance on local authorities in the UK concluded that:

> While external triggers to management change are important, the susceptibility of individual authorities to change, and the direction of that change, is related to internal power relations and to local sensibilities and circumstances.
>
> (Lowndes 1999: 37)

It can take a long time to change traditional ways of working; new ideas merge with existing structures and institutions to produce 'different, and sometimes contradictory, streams of ideas and practices' (Lowndes 1999: 37). This means it is important to stress historical context and its local effects when considering the development of quality of life indicators. Rhodes (1997) stresses the importance of ethnographic research (as employed in my case study), as a means of capturing attitudes towards participation; these will necessarily vary according to institutional norms and working practices. Participation remains a radical idea for some, who are unsure of its merits and therefore reluctant to commit the time and resources it needs.

Fraser (2005) has developed a typology of participation approaches. Although based on individual policy actors rather than an analysis of institutional discourses it is useful for thinking about the latter. She identifies four main types of attitude to participation and their respective community constructs which I have condensed in Box 4.1.

If we want to talk about power we have to think about 'who is constituted as "the community" and how the community's interests are understood' (Fraser 2005: 287). Scholars have become increasingly wary of how central government discourses deploy the notion of community participation and have used governmentality theory to analyse this (for example Taylor 2007; Raco 2005, 2007; Flint 2002). Liberal governance seeks to encourage and foster autonomy and individual freedom but creates a management problem in the process. The theory of governmentality (Foucault 1991a) (and the clue is in the word) refers to the way that governments in liberal democracies manage large populations by influencing not only opinions and attitudes but also, more profoundly, how we think about ourselves as humans in the world. Power operates therefore not by suppressing autonomy or freedom but by creating particular preferences, or subjectivities (Rose 1996). The idea of the self is influenced by the way dominant discourses create particular truths; this is how the state governs an otherwise unmanageable population. Some characteristics of the 'governmentality framework' include: controlling at a distance through networks; the 'recasting of subjectivities'; the construction of objects, ideas and entities (such as quality of life, sustainable development, communities, participation); and techniques that manage and direct reality (such as indicators, community regeneration strategies, consultation procedures) (adapted from Rydin 2007: 611). Foucauldian analyses of New Labour community participation discourses point out how communities are constituted and how subjectivities of 'active citizens' are created to respond to central government notions of 'sustainable communities'. These are complicated and entwined discourses constructed both from historical notions and new ideas. This complex is now discussed.

The entwined discourses of community, locality and disadvantage

Etzioni defines community as a social entity 'resembling extended families' (not necessarily geographically based) which is 'normative-affective', that is, which shares (and shapes) moral norms, and where webs of affection exist between members (Etzioni 2007: 24; 2001). However, for New Labour 'community' was a locality-based concept where neighbourhood communities had latent values that government initiatives could revive (Fremeaux 2005). Making the geographic

Box 4.1 Typology of approaches to participation (adapted from Fraser 2005)

Anti-/reluctant communitarians and economic conservative approaches

Community is constructed as 'mythical' or 'sentimental' and actors argue for non-state intervention and self-regulation of citizens while often (paradoxically) arguing that the community will look after the individual. Actors acknowledge the need for community support but that this is not really 'work' and can be largely done through voluntary work. Communities who do not contribute to economic agendas are ignored.

Technical-functionalist communitarians and managerialist approaches

Community is constructed as a relatively stable and homogenous entity. Actors see community engagement as 'maintaining equilibrium'. Their goals are a minimum of fuss, maximum efficiency and they rely on expert-driven consultation with communities. They see participation as apolitical and participants are often recruited from well-established community groups for their abilities. Actors can see themselves as neutral arbiters of disputes. Under this approach conflict is avoided and notions of justice largely ignored.

Progressive communitarians and empowerment approaches

Community is constructed as complex and problematic. Actors see social justice as important and they pay attention to the processes of participation. They see participation as a change agent but their emphasis is on incremental rather than radical change. However, while focusing on the politics of inclusion/exclusion, wider structural impacts on communities are largely ignored. Actors are generally egalitarian and inclusive in approach, relying on face-to-face contact and debate. Conflict is acknowledged, while not necessarily ensuring that under-represented groups are present.

Radical/activist communitarians and transformative approaches

Community is constructed as esteemed places where 'ordinary folk' live and real life takes place. Actors are concerned with discrimination and oppression. They link personal and local issues to national and global ones, seeking to transform social order. Their focus is on the redistribution of resources and the fight against poverty. Power relations are at the forefront of their analysis of problems. Many see existing community participation as a smokescreen to the real issues of injustice. They seek to recruit people who are often sidelined and prefer bottom-up approaches.

community the unit for decision-making and participation is problematic in that people may identify less with their neighbourhood than with associations formed through interests, workplace, or internet-based social networks. There can also be issues of deep-seated social conflict within local groups, which may therefore be 'weakly organised, fragile and vulnerable to manipulation' (Buckingham-Hatfield and Evans 1996b: 12). Empirical research reports that neighbourhood-based programmes can be 'insensitive' to cultural differences within communities, meaning that some people are less able to participate than others (Dinham 2006). There may also be pressure from central government for local councillors to represent their areas with 'one coherent voice, resolving conflict and producing consensus which can then be presented, neatly packaged, as the local view to the centre' (Barnett 2011: 285). New Labour used words such as 'empowerment' and 'inclusive' with enthusiasm but with insufficient critical engagement with questions of how power relations actually work within communities (Wright *et al.* 2006).

For New Labour 'sustainable communities' was a key concept in its attempt to mainstream sustainable development and modernise local government, and represented 'the essential building blocks of social harmony and progress' (Raco 2007). This construction of an ideal locality-based community underpinned work on national quality of life indicators that built on the idea of 'New Localism' in Local Agenda 21. This argues that local policy initiatives will deal more effectively with contemporary ecological problems by creating 'a more rational future with local government leading the development of more sustainable communities, life and work styles' (Marvin and Guy 1997: 311). However, this focus on the local can create a 'black box' where communities become disconnected from the national and global context and where wider structural impacts on the local are ignored (Marvin and Guy 1997: 312). Problems are then identified at the level of community and the community also becomes the instrument of solving them. This construct of community as both problem and solution was a key feature of New Labour discourses. However, there is an increasing awareness of 'wicked problems', whose solution lies both within and outwith the scope of neighbourhood, and where the perpetrators of and those affected by environmental problems do not fit neatly into administrative boundaries (Hajer 2003; Barnett 2011). Increasingly people may find more effective arenas for deliberation and participation than their local community or area. For instance, the rise of internet activism or 'clicktivism' gives people easy access to have their say on environmental and structural issues outside the scope of their local arena.

These discourses also hold implicit messages not only about what sort of community should be developed but also what sort of people should make up that community (Raco 2005). These are reflected in AC guidelines for developing quality of life indicators. Labour's understanding of participation sought to create a particular type of citizen and it 'fail[ed] to meet people on their own terms' (Wright *et al.* 2006: 348). As set out earlier, New Labour's policies concerning participation and social capital focused particularly on 'disadvantaged' neighbourhoods where residents were 'subject to a high level of demand for participation, community involvement, and representation from local government and agencies' (Jupp 2007: 2834). Historically, participatory practices have been located around the empowerment of disadvantaged communities. Fremeaux argues that the idea of community gained prominence during the 1960s when high-rise blocks were criticised for their soullessness and lack of community spirit. The term therefore became associated with a particular set of meanings:

> Community became the answer to the predicament of the poor and underprivileged, a connotation which has remained ever since: since the late 1960s policy-makers have consistently used the term to refer to the socially excluded.
>
> (Fremeaux 2005: 267)

Participation is often imposed indiscriminately on the poor with little critique of its value, appropriateness or effectiveness. For example, Cooke (2001) points to the wealth of social psychology literature which has highlighted the problems of group decision-making but which has not informed participatory practice in relation to the poor:

> ...the poor of the world, particularly but not exclusively those in 'developing countries', are the victims of a disciplinary bias: put simply the rich get social psychology, the poor get participatory development.
>
> (Cooke 2001: 121)

Disadvantaged communities have often been the target of participation strategies but without incentives members may find it harder to attend regular meetings:

> Despite its egalitarian overtones, 'participation' is always contingent on some 'trade off', whereby, clearly, some have a lot more to trade with than others.
>
> (Bush 2005: 27 citing Peterson and Lupton 2000: 162)

By making the concept of active citizenship central to the solutions of social exclusion, the discourse limits the legitimisation of alternative accounts of the causes of deprivation, for example national economic policies. This is an inward-looking approach to the problems of communities: the solution is to enhance the individual's ability to be politically and economically engaged. Individuals are 'empowered as agents' to solve the problems of poverty in their neighbourhood and the 'hierarchical structures of national and global are sidelined' (Wright *et al.* 2006: 355).

In addition, possibilities for participation may be limited by requirements to use it to meet central government agendas such as developing local strategies to tackle crime and anti-social behaviour, filtered through various government-controlled mechanisms and evaluated by a range of central government instruments that steer decisions in a particular direction (Wright *et al.* 2006; Jupp 2007). The government decides 'how the community will be involved, why they will be involved, what they will do and how they will do it' (Wright *et al.* 2006: 349).

The NDC also demanded a 'what works' evidence-based approach. This is problematic for two reasons. First, the evaluation requires highly complex skills most commonly found in the academic and research fields. This type of knowledge is privileged over alternative forms. Local residents with ideas about what would work in their own areas find themselves at a disadvantage if they cannot 'evidence' them (Wright *et al.* 2006: 356). Jupp (2007) found that greater legitimisation was accorded to participants who had learnt to describe their communities as 'deprived' in the language of regeneration policy discourse and who could produce the sort of knowledge that was needed. Therefore, participation led to a form of knowledge production that, while appearing to facilitate empowerment, actually upheld central agendas (Jupp 2007: 2832).

All these accounts express concerns about how power is exercised and by whom. This is partly because in New Labour discourses on the concept of community dominated the notion of participation and partly because New Labour was seriously conflicted between centralist and localist tendencies. Participation was encouraged as a means to create sustainable communities, and could only happen within these (neighbourhood) structures. The concept of 'community' focuses on the 'disadvantaged', and the increase of social capital is seen as a way for them to help themselves out of poverty. Etzioni asks, now that women are more occupied by paid work, who will do this community service? He turns to the young, the elderly, the unemployed in addition to assuming that those women who are around will continue to do the work (Etzioni 2001). This dovetails with SD discourses on the greater participation of the vulnerable, disadvantaged and marginalised, but whereas SD discourses stress social justice and empowerment for citizens, Third Way discourses emphasise responsibility for the social good, duty and economic efficiency; responsibilities which lie largely with disadvantaged communities and particularly their more vulnerable members. In SD discourses, citizens have the *right* to participate, but in LGMA discourses they are increasingly *expected* to participate through a 'responsibilisation' of communities. In the LGMA this is part of a wider attempt to create an 'activation' of citizens who take responsibility for their lives and depend less on the welfare state. However, this polarisation can be too simplistic as the LGMA and SD discourses are intertwined:

> Community empowerment represents a central element in the rolling out of neoliberal relations and subjectivities and acts as a mechanism for instilling the illusion of self-sufficiency and societal detachment. At the same time, community plays a central role in the discourses of SD…Within SD discourses a sustainable community is one whose collective resources, such as employment and forms of collective consumption, are maintained in the longer term to enable community change to proceed in less rapid, more equitable ways…the approaches do have overlaps, but their contrasting interpretations of the processes and practices of development planning necessitate further empirical analysis.
>
> (Raco 2005: 331)

In these discussions the local authority holds an ambivalent position, being both part of the state and representing local people to the state, including complaints against itself (Geddes 1996; Barnett 2011). Local councillors also become caught between their representative function (member of a political party) and their role as community champion for all residents (Barnett 2011), and therefore become implicated in changing discourses about their role. It is therefore important to look closely at local governance processes and discourses. Major reviews of the key instruments of the LGMA and SD mainstreaming point to several problematic issues. Geddes *et al.* (2007) report a 'wide local variation' in the progress of LSPs throughout the country, identifying both 'virtuous' and 'vicious' circles of development. Virtuous circles are where 'LSPs become deeply embedded in the local governance landscape as sustainable institutions and conform broadly with the aspirations for partnership working in New Labour philosophy'. Vicious circles are 'when governmental aspirations are thwarted and a destructive cycle ensues, representing a challenge perhaps not only to LSPs but to the partnership mode of governance *per se*' (Geddes *et al.* 2007: 110). The main factors which create a virtuous circle are established working relationships and trust among members; a dedicated staff team with resources and support; a stable political environment; the wide ownership of the community strategy; clear strategic goals; and good publicity to ensure that the LSP is recognised in the wider community. The lack of these factors can lead to a 'limited and superficial' engagement, with strategies being developed by a few people and just 'signed off' by the LSP. Geddes *et al.* (2007: 113) highlight 'fundamental issues about the concept and practice of "partnership governance" – from the capacity of partnerships as institutions with wide remits but limited resources, to their ambiguous relationship to local democratic accountability'. There were significant problems about involving stakeholders in the process of developing the community strategy and creating a sense of ownership within a wider partnership (Darlow *et al.* 2007).

Conclusion

With particular relevance to the case study to follow, there is a rich literature exploring the UK shift towards local community governance in New Labour reforms which finds that the reality was far behind the rhetoric. New Labour implemented a range of policies to increase community capacity to 'contribute to democratic policymaking, to deliver mainstream services and to build social capital' (Taylor 2007: 298). Yet participation was often heavily prescribed in accordance with central agendas, and the ability for local people to influence decision-making was limited to the extent that some argue that the move towards community governance in the UK was a new form of state control where language played an increasingly key part (Taylor 2007; Jupp 2007; Wright *et al.* 2006; Fairclough 2003). The literature on participatory discourses and practices in the UK is at best circumspect and at worst deeply cynical.

Therefore, in any discussion of participation a 'critical understanding of terms' is needed (Fraser 2005: 288). State encouragement of participation may aim to decentralise government, cut costs to government, transfer rights or responsibilities, or subdue opposition, so it is important to ask what the government is asking communities to do and why (Wright *et al.* 2006). Scholars have pointed out that the tendency in participatory practices to focus on consensus rather than difference can lead to reductionist and over-simplistic accounts (Cahill, 2007; Laessoe 2007; Kothari 2001). Rather, as Cahill (2007: 2862) suggests, the 'struggle between multiple perspectives' can be understood as a 'productive contestation' and the 'hallmark of democratic traditions'.

Democratic engagement is a much wider and different concept than community participation. Linking participation to neighbourhood limits the scope for the legitimisation of discourses which identify wider structural processes as a barrier to wellbeing. Amin (2005: 628) criticises this 'localisation of the social' which New Labour community participation discourses promoted and argues instead for a 'public culture of the commons' where the focus is on 'an ethic of empathy or, at minimum, tolerance for diversity and difference'. Rydin, in her critique of environmental planning processes, suggests that, due to resource implications and the need for action, deliberation and collaboration with local groups should be reserved for particular places in the policymaking process where they can be most effective. Furthermore, she argues that these forums could be used to go beyond debating particular planning decisions, for example, to more fundamental issues like the morals and ethics which should underpin planning. She argues that 'the most profound issues' should be handed back to 'the democratic realm' (Rydin 2003: 182). Pateman's (1970) thesis for a society-wide culture of participatory democracy which is linked to economic stability and equality and which runs through all aspects of our lives, still resonates today. The current discourse coalitions of subjective wellbeing, social capital and community participation which focus heavily on neighbourhood place-making, present very 'thin' versions of participation and empowerment. There is rarely any detailed engagement with local power processes. Furthermore, local government may be unable to deliver 'empowerment' because this means 'tackling fundamental inequalities' (Barnett 2011: 286). While not seeking to devalue the many innovative and excellent local initiatives which have made a positive difference to residents, these alone cannot tackle deeper issues which affect society. At worst, over-emphasis on local wellbeing projects may distract from debates about the wider and deeper issues by making local residents responsible for delivering their own wellbeing.

What does all this mean for the inclusion of ideals of participation within indicator sets of wellbeing? Or for the process of developing indicators through participative processes? How would we include a power sensitive approach to looking at participation? To illustrate some of the issues in this chapter I offer in Table 4.1 extracts from three very different accounts of wellbeing (discussed already) which make an interesting comparison in terms of what they have to say about participation. Although direct comparison is difficult as they all use different constructs of wellbeing, it is possible to get a flavour of the different emphasis that different accounts put on participation and (in the case of the

Table 4.1 Comparison of participation constructs in three different accounts of wellbeing

Martha Nussbaum 2000 *Central Human Capabilities* Capability approach	*European Network on* *Indicators of Social Quality* Social quality	*The Audit Commission 2005* *Local Quality of Life Indicators* Sustainable communities
• Being able to participate effectively in political choices that govern one's life; having the right of political participation, protections of free speech and association. • Being able to live with and toward others, to recognise and show concern for other human beings, to engage in various forms of social interaction; to be able to imagine the situation of another. (Protecting this capability means protecting institutions that constitute and nourish such forms of affiliation, and also protecting the freedom of assembly and political speech.) • Being able to use one's mind in ways protected by guarantees of freedom of expression with respect to both political and artistic speech, and freedom of religious exercise.	• Trust in government; elected representatives; political parties. • Importance of politics. • Volunteering: number of hours per week. • Willingness to pay more taxes if it would improve the situation of the poor; the elderly. • Willingness to do something practical for the people in your community/ neighbourhood. • Membership (active or inactive) of political organisations. • Proportion of residents with citizenship. • Proportion having the right to vote in local elections and proportion exercising it. • Proportion of ethnic minority groups and women elected or appointed to parliament. • Percentage of labour force that is a member of a trade union (differentiated to public and private employees). • Existence of processes of consultation and direct democracy (e.g. referenda). • Number of instances of public involvement in major economic decision-making. • Percentage of organisations/institutions with work councils. • Percentage of the national and local; public budget that is reserved for voluntary, not-for-profit citizenship initiatives. • Marches and demonstrations banned in the last 12 months as proportion of total marches and demonstrations (held and banned).	• The percentage of people surveyed who feel they can influence decisions affecting their local area (aspirational indicator). • The percentage of residents who think that for their local area, over the past three years, community activities have got better or stayed the same. • Election turnout.

Sources: Nussbaum 2000; Phillips 2006; AC 2005.

ENIQ and AC account) how they seek to measure it. Do the indicators in the ENIQ or AC indicator sets reflect the ideals set out by Martha Nussbaum's list of central human capabilities? What does Martha Nussbaum highlight or leave out? What do the indicators actually tell us? What can't they tell us? We should also bear in mind that the AC set is a set of 45 indicators focusing on local community quality of life, and the ENIQ set comprises 95 indicators focusing on national societal quality.

5 The role of indicators

Knowledge, rationality and public policy

On 26 January 1982 Margaret Thatcher was not happy. Unemployment had reached three million for the first time since the 1930s. This hit the UK national headlines on that day and caused the prime minister a great amount of difficulty in the House of Commons (BBC News 1982). The figure was a powerful indicator, (un)popularised by the tabloid term 'Maggie's millions', and was successful in that it put pressure on the government to bring about change. However, the methods for calculating unemployment changed thirty times between 1979 and 1998 (MacGillivray 1998). Governments have been criticised for creating strategies to obscure the actual levels of unemployment, and technical and political arguments about the proper way to obtain a full picture proliferate (for example, Beatty and Fothergill 2005). In her studies of social indicators in the USA, Innes (1990) points out that long before any reliable measure of employment existed, Congress, concerned about rising unemployment, simply *voted* on a suitably high number to capture the attention of the nation. This was successful, she argues, as the indicator was already 'meshed with publicly understood concepts' and it became quickly 'institutionalised' (Innes 1990: 2). The most influential indicators are not necessarily successful due to their 'objective' or 'scientific' status, although this can be crucial at times depending on their purpose and context. Their power to influence is greatly enhanced if they already matter to the public, if they resonate with people's practical judgement, are easily communicated and publicised, cause debate and make politicians act. As many have argued, if an indicator is to be linked to policy action it must indicate an issue that already means something to us. Statistics and indicators can and do play a important role in creating and changing this meaning, but they do so within an inextricable relationship with politics, culture and context and this relationship is often ignored to the detriment of effective policy creation.

The power of the 1982 unemployment indicator lay in the meanings it held for society and its dramatisation by opposition parties and the media. There is no single objectively derived number where unemployment figures suddenly show mass unemployment and transform the country into a pre-war society. Three million is a threshold which has been invested with meaning and through which society's fears and concerns can be mobilised. Unemployment has a real and profound effect on people's lives and it matters how we assess it. Employment is a major component of quality of life for most people. It means having access to resources, to purpose and structure, to self-worth and social contact. However, we also, individually and as a society, invest unemployment with various meanings; the unemployment of individuals does not necessarily reflect failure of government, because they may be, and often are, seen to be reluctant to seek work; mass unemployment, on the other hand, signals government failure. It is seen as a major problem for society, not just for those individuals out of work. Increased crime, breakdown of communities and an overall reduction in life chances for young people are among the social ills commonly attributed to high levels of unemployment.

Indicators such as those for GDP and unemployment are extremely powerful because they generate debate and influence attitudes and policy. However, Boulanger (2007: 15) asserts that there is a 'widespread feeling of deception' because sustainability indicators have failed to be politically influential or universally accepted. A European-wide study of indicators, the PASTILLE Project, using case studies within local authorities and municipalities, echoes this concern:

> Although indicator-type tools are seen as useful instruments for measuring and communicating the complex goals of urban sustainability, the policy actors are, broadly speaking, disappointed with their effectiveness.
>
> (Astleithner and Hamedinger 2003)

As outlined in Chapter 1, local indicator sets are intended to provide a framework against which local policies and practices can be created and evaluated. These policies should be underpinned by whatever construction of the good life local residents have determined.[1] The development of indicators can be a lengthy process, particularly when involving local communities (Lingayah and Sommer 2001; Sommer 2000). Further time is needed before those indicators become politically influential (Innes 1990). Policymaking in itself is a complex and uncertain process, involving many actors and is subject to a range of pressures and forces (Kingdon 1995). Taking all these things into account, international research has found it hard to make direct links between the development of sustainability indicators and policy or practice change in the governing body concerned (Boulanger 2007; Rydin *et al.* 2003).

The literature offers many reasons why indicators might be ineffective; whether explicitly stated or not, these are inevitably linked to theories about how policy is made and the role of knowledge in policymaking. Depending on which theories inform these discussions, different reasons will be given for indicator failure or success. Indeed, different constructions of failure and success will be used. This chapter will review two dominant and contrasting paradigms of policymaking commonly presented in the literature on indicators: the 'rational' and 'discursive' models. I will consider the discussions of indicator effectiveness in relation to them.

In this chapter I look at the role of scientific discourse and power, and ask which forms of knowledge are legitimatised within policymaking processes. This builds on my commentary in the previous chapter regarding the role of knowledge in terms of participation. I will set out the national policy context on indicators and also draw some lessons from a well-documented London community indicator project. This will lead to a discussion of the methodology of a case study, involving collaborations between academics and practitioners and the tensions and potentials inherent in this relationship. I will end the chapter by drawing together the arguments made previously before moving to Part II of the book.

Indicators and policy process

The fundamental requirement for a successful and influential indicator is that it must reflect something that is already important to society:

> The things that get measured should evoke happiness when they are improving and unhappiness when they are getting worse – if the change doesn't matter to the community then you are not measuring the right thing.
>
> (Lawrence 1998: 80)

Other attributes, such as scientific legitimacy, help but they tend to come into play only when the indicator reflects an issue that has already attracted policy attention. Only a few issues are dominant in public discourse at any one time, and an indicator's success in securing policy recognition is increased by being legitimised by a wide range of people in public arenas (Blumer 1971; Innes 1990; Boulanger 2007; Rey-Valette *et al.* 2007). Kingdon's (1995) work shows that the agenda-setting process is not linear but involves independent streams (policy, problems and politics) which have their own pattern of action and reaction according to different forces. At least two streams must join in order to create a policy window where an issue can be recognised and get onto the agenda. He argued that policymakers do not evaluate different options systematically and that policymaking is more fluid and unpredictable than commonly thought. This built on the earlier well-cited 'garbage can model' of policymaking by Cohen *et al.* (1972) who argued that many institutions were 'organised anarchies' where a complex relationship of four streams (problems, solutions, choices opportunities and participants) existed, the situation being magnified when communication was needed between different departments of an organisation.

A set of indicators to measure and subsequently change an issue that is not widely legitimised, of public concern or high on the policy agenda (such as ecological sustainability) must compete with well-established indicators in an uncertain policy process where only a small number of issues gain recognition. This is a major challenge for indicator sets built on theories of sustainability or wellbeing, theories which seek to challenge the dominant economic discourses. In order to alter discourses, the indicator process must work through and on them and through and on those institutions which promote them. That is why this book places such a strong emphasis on the institutional *discursive process* of developing indicators *within* the policymaking arena.

The rational model: policy without politics

This model of policymaking can be described as rational, scientific or technocratic (Boulanger 2007; Innes 1990; Dery 2000). Here decisions are made on rational grounds using objective, scientific evidence, and policymakers are responsive to new evidence, changing their preferences and goals accordingly (Dery 2000). They choose between alternative options to solve specific issues based on efficiency and effectiveness, using techniques such as cost/benefit analysis and statistical modelling (Boulanger 2007; Wagle 2000). This positivist approach is grounded in a theory that knowledge is created by unbiased experts outside the political process and that this evidence can be used by policymakers for decisions (Innes 1990). Boulanger (2007: 17) argues that this 'erects a firewall between science and politics, between the rigorous world of facts and logic on the one hand, and the subjective world of values, ideals, beliefs on the other'. This approach tends to reduce policymaking to a problem-solving exercise that is 'technical, rational and transferable' (Rydin *et al.* 2003: 4). Although this can serve to oversimplify or misrepresent problems and solutions, it is still how the majority of people view the role of knowledge in public policy (Turnhout *et al.* 2007; Rydin *et al.* 2003; Wagle 2000; Innes 1990).

> The belief in the ability to apply knowledge and to manage situations with a specific goal in mind is *the* modernist belief and it is one which has proved slow to recede even in apparently post modern times.
>
> (Rydin *et al.* 2003: 546 [original emphasis])

Astleithner *et al.* (2004) in their work on local sustainability indicators argue that the development of indicators is seen by many policymakers as part of a rational evidence-based policymaking exercise. They see indicators as 'non-subjective tools', 'decision support instruments' and 'exogenous to the process of policymaking'. Under this paradigm, indicators tend to be judged on their scientific, technical and economic properties (Boulanger 2007). Indicators are regarded as instruments, made outside the policymaking process to be imported and used by policymakers as appropriate. Guidance for creating indicators often reflects this paradigm, stressing that indicators need to be 'measurable according to available data', 'scientifically robust' and 'comparable' (PASTILLE 2002: 11).

The discursive model: policy as politics

This model rejects the division between science and politics and sees policymaking as influenced by and responsive to various discourses in 'a struggle over the definition, explanation and interpretation of public problems' (Boulanger 2007: 18). Wagle argues that under this paradigm, the 'understanding of subjective values and social systems' is an integral part of policymaking leading to better policy due to the greater understanding of which policies may impact on who, how they impact and why. She argues that policy is driven by 'arguments and subjective values' rather than through the application of scientific theories (Wagle 2000: 210). Therefore the rational view, which still predominates in practice, that 'experts' should deliver robust and valid data divorced from and prior to policymaking, is problematic and naive.

This discursive view of the role of knowledge in policy formation is more context-based and complex than the scientific view (Innes 1990). Healy (2001: 2) argues for a communicative approach to public policy around the promotion of sustainability to create understanding and progress 'by exploring the complex social processes through which meanings are created'. This model takes a broader view of what counts as knowledge and the ways in which that knowledge can influence without being actively or directly used. It suggests a two-way relationship between indicator development and policymaking:

> The most influential, valid, and reliable social indicators are constructed not just through the efforts of technicians, but also through the vision and understandings of the other participants in the policy process. Influential indicators reflect socially shared meanings and policy purposes as well as respected technical methodology. If they were not simultaneously technical and political creations...they would not be valid, since the very concept of validity implies a correspondence of measure and meaning.
>
> (Innes 1990: 4)

This model sees a wide variety of roles for indicators including justifying or manipulating policy discourses (Rey-Vallette *et al.* 2007). Although often these discourses lean heavily on 'facts' and 'scientific' knowledge, it is the political context that determines which facts will be relevant (Boulanger 2007).

The discursive view of policymaking lays more stress on the democratic and deliberative processes of creating and using indicators, and less on the scientific or technical legitimacy of the indicators themselves (although the latter is not seen as unimportant). Indicators are both inputs and products of policy discourse. They not only measure an issue but also construct it through discussion of its properties, victims, guilty parties and solutions, and this is dependent on the values of those discussing the issue (Zittoun 2006). Indicators build the issue as much as they are built by it. In this paradigm therefore, characteristics of indicators likely to lead to success are that they are easy to understand, reflect social importance, are widely legitimised, and can compete with other issues for policy attention.

The fact is that both rational and discursive elements can co-exist in policymaking. However, the rational model dominates public policy discourse, and the effect can be to privilege 'rational' expert-driven evidence over other types of knowledge and over normative values in the development of indicators. It can also inhibit the creation of spaces for collaborative enquiry into what we value and what we want to measure, and mask the reality that policy is not driven by evidence but by a meshing of contextual knowledge, political values and cultural context which are in a complicated relationship with 'evidence'. If policymaking is believed to proceed rationally through the application of 'objective evidence' then discussions of power can be side-stepped. Although participatory processes are moving into the mainstream, the rational expert-led approach to policymaking remains resilient. In the foreword to a report on happiness economics commissioned by the Westminster-based Institute of Economics, a director of that organisation states:

> There is no question that happiness data is being used selectively to justify preconceived beliefs about policy alternatives.
>
> (Johns and Ormerod 2007: 9)

What is striking about this statement is not the revelation itself but the fact that it is treated as such. Social scientists have long known that knowledge and science in policymaking are mediated through power struggles over political values and I set out some of that research here. Ironically, in the face of extensive research which shows otherwise, we persist in believing that rational evidence-based policymaking is the norm and view anything other than this as suspect. How can we properly assess the success or failure of policy if we continually fail to engage with the realities of rationality and power? (Flyvbjerg 1998). Even when politicians acknowledge the role of values in politics, the role of statistics is seriously over stated. Take this interesting statement from David Cameron in a question and answer session about wellbeing measurement:

> Often what politicians do is seek evidence to back up the view they already hold, and so it is one of the reasons we have an independent statistics office, to try and stop us from doing this.
>
> (Cameron 2010)

This is a circular 'problem', unless of course there is one 'true' set of evidence. I show below, in terms of wellbeing measurement that is increasingly unlikely.

Scientific rationality and indicators

As Foucault (1991b) argued, science is one of the main legitimising discourses in our society for truth claims, so scientific legitimacy is important in indicator development. Science is obviously important as well, because it tells us things we need to know about an issue in order to act on it and it can tell us if our actions have made a difference. But this is not the same as suggesting that science is the *only* thing that is important, or that there is only one way of measuring something, or that science drives indicator use and development. The work of Collingridge and Reeve argues that a reliance on science alone to provide the answers we seek paradoxically undermines it:

> The aim was to limit political dispute by appeal to objective, scientifically established facts which no-one could deny without losing all credibility as a rational agent, but instead the political debate has been widened by generating

a technical argument about the data and its interpretation. As more scientific research is done in the hope of limiting the arguments, the reverse in fact occurs since there are an increasing number of technical issues under dispute.

(Collingridge and Reeve 1986 cited in Turnhout *et al.* 2007: 222)

Even at the very 'scientific' end of the sustainability-measuring spectrum, in dealing with ecological indicators, Turnhout *et al.* (2007: 221) show how science and policy 'enter in some kind of joint knowledge production' and that indicators can only be evaluated on a 'case-by-case basis' according to their context.

This misconception that science will provide an objective answer for policy is well illustrated in the literature on indicators. As the technical sophistication of measurement instruments is increased, with the aim of getting nearer to the truth of sustainability or wellbeing, a wider choice of measurements complicates policymaking. Ortega-Cerdà (2005) shows how various organisations generating sustainability indicators rely on appeals to objective methods as a source of credibility and legitimacy. However, by comparing two different sets of indicators, which had shared the aim of measuring the sustainability of different countries, he shows how they obtained almost opposite results. This dilemma is also well illustrated by researchers working on the national Canadian Index of Wellbeing. They identify 21 'critical issues' to consider when constructing wellbeing indices, including: how to apply temporal and spatial coordinates, how to categorise groups, whether to use objective or subjective measurements. They calculate that even if there were only two alternatives for each of the 21 issues (a conservative estimate) over 2 million different sets of indicators might be constructed (Michalos 2011: 12).

In addition to this increasing complexity, and possibly because of it, policymakers will tend to use the information that suits their policy preferences and to reject evidence that points in a direction they are not prepared to go (Tenenbaum and Wildavsky 1984). Boulanger (2007: 18) argues that 'political controversies are immune to resolution by appeal to the facts because the conflicting parties simply disagree on which set of facts is to be considered relevant'.

Despite the overwhelming evidence that statistics are not only highly contested but also heavily manipulated, the UK's national statistician, on discussing the new national initiative to measure wellbeing, claimed that 'statistics are the bedrock of democracy' (ONS 2010). When statistics are elevated to the same status as free speech, justice and equality, I suggest we are in trouble.

We should fundamentally challenge the tendency to adopt 'objective' and 'scientific' enquiry as unproblematic sources of the 'truth'. Rather we should be asking what forms of knowledge and experience are undermined and which accounts of the world are privileged (Foucault 1991b). Foucault set out three types of knowledge: the dominant knowledge, based in scientific theory; historical knowledge (which may be elite or expert knowledge but which has been 'buried and disguised' because it did not conform to dominant discourses); and local or popular knowledge which is dismissed as inadequate to the task in hand. He regarded these last two as 'subjugated knowledge', disregarded by dominant scientific enquiry, and argued for a 'return of knowledge'. Foucault offered no analysis of what this knowledge could or should do but acknowledged its 'claims to attention' and asked us to be aware of what its recognition might bring to society (1991b: 78).

As already set out, these issues of how knowledge and power relate to each other are crucial to indicator development. Zittoun (2006) argues that indicator development processes reveal the 'complex relationships which develop between knowledge, expertise and power' (cited in Rey-Valette *et al.*, 2007). Wagle (2000: 210) claims that policymakers attempt to keep the capacity for policymaking in their own hands by 'propagandizing that non-experts have first to gain the expertise to engage themselves in policymaking process'. This has been found by many empirical studies of community participation as mentioned in the previous chapter. In terms of policymaking and policy change the notion of power is vital because policy development can be seen in terms of 'changes in the configuration of dominant interests', where policy innovations result from a combination of power and ideas (Dery 2000: 39). Similarly, Boulanger (2007) shows how indicators can be important tools in the struggle for power and influence.

> What is needed, therefore, is a paradigm shift in the whole concept of policy making by reducing the hegemony of expert-knowledge and by adapting discourse-oriented processes.
>
> (Wagle 2000: 212)

Some scholars may see indicator development as 'control at a distance' due to the danger that indicator development at a local level is responding to national discourses around sustainable communities (as discussed in the last chapter). In this way indicators can be seen as instruments of governmentality. More discursive approaches to indicator development that value, include and legitimise different types of knowledge, may have the power to resist such tendencies (Rydin 2007). Much literature talks about the need for expert and lay knowledge to come together (for example McAlpine and Birnie 2005; Turnhout *et al.* 2007). In this way the development of local quality of life indicators (QLIs) in recognising the role of local knowledge and values can be seen as a move from a 'technocratic mode of looking at SD towards SD as a social construct' (Bayliss and Walker 1996: 87). Levett argues for a dialectical rather than deductive process of indicator development:

> The struggle to frame measures leads us to a better understanding of what we are concerned about. Grappling with the question 'how can we tell whether it is growing better or worse?' both forces us and helps us to clarify what we think 'it' is. Indicators are not neutral technical entities: they are inescapably value laden. They are inputs into policy as well as consequences of it.
>
> (Levett 1998: 294)

National contextualisation of indicators

Local indicator sets are influenced by national-level sets of indicators and also by central government guidelines of what local indicators should look like. They are also set within a culture of increased performance measuring under NPM models of governance (PASTILLE 2002). Power (1994: 13) criticises the NPM model for having created an 'audit explosion' and argues that systems of auditing support strong central control whilst appearing to devolve power:

> These competing pressures, to devolve on the one hand and to control on the other, constitute a distinctive idea of government. Consistent with a liberal mission, the UK state is increasingly committed not to interfere or engage in service provision directly; it seeks to fulfil its role by more indirect supervisory means. In many cases the state has become regulator of last resort, operating indirectly through new forms of control …which have the appearance of being apolitical

This section briefly reviews the main system of central auditing for local authorities in the UK during the New Labour government and discusses what impact this may have had on the effectiveness of QLIs to find policy space. Following the 1999 Local Government Act the mechanism of Best Value was introduced to replace CCT. The emphasis was on 'economy, efficiency and effectiveness' (UK Government 1999). The Best Value Performance Framework placed a duty on local authorities to deliver services according to centrally set standards. It was the main mechanism by which central government held local government to account. One of the requirements of this auditing was that councils had to report annually on a set of 90 Best Value Performance Indicators (BVPIs). Each council would have individual targets set alongside each of these indicators. Comprehensive performance assessment (CPA) was a new regime attached to Best Value in 2002 which started to introduce a reward and ranking system for local authorities. Councils were graded 'excellent', 'good', 'poor' and 'worst' according to the assessment regime, and rewards were in the form of an 'inspection holiday' whereas 'worst' councils could expect increased auditing.

The Audit Commission (AC) oversaw this auditing process. It is also the body that researched and promoted quality of life indicators. Because of the AC's role of overseeing local government performance, it tended to promote QLIs as a performance management tool rather than as a means to support a collaborative process aimed at creating greater understanding of and action towards sustainable development aspirations (Shepherd 2007). The AC provided a set of QLIs for local authorities to use (if they wished) but there were problems associated with providing such an off-the-shelf solution in that local authorities might side-step developing QLIs through participatory methods with the LSP. Research also suggests that in treating the indicators as tools for performance management rather than as instruments for community development, local authorities may direct their attention 'upwards towards government rather than outwards to local communities' (Boyne 1999: 4). Reporting on a major review of community strategies and their capacity to deliver 'joined-up governance', Darlow *et al.* (2007) note that there was little evidence of joined-up service delivery on the ground. Although community strategies have helped joined-up thinking within local authorities, they have had limited success in stimulating joined-up working across organisations. In addition, there was a 'wide variation in the form and content' of the strategies considered which reflected a 'fundamental tension within community strategies, in particular, the extent to which they provide a means of accountability or performance management mechanism between central, regional and local government' (Darlow *et al.* 2007: 127). These local variations and difficulties are important factors in considering the particular institutional contexts within which local indicators are created and used.

The pressure on local government to be accountable must be considered when developing local indicators and determining under what conditions those indicators are most likely to be effective. A key observation from a detailed study of indicator use in local policy was that indicators which made an impact on policy were those for which there was already a statutory responsibility to report on – i.e. where that indicator was also a BVPI (Shepherd 2007). This raises questions about the potential effectiveness of non-mandatory indicators which reflect other aspirations.

However, in 2006, Ruth Kelly the Secretary of State for Communities and Local Government, announced the creation of a 'Lifting the Burdens' taskforce to review the bureaucratic loads on local government. As a result, BVPIs, and other sets of indicators, were abolished, and the only indicators which central government required local authorities to report on were a national set of 198 indicators. CPA was replaced by comprehensive area assessment which promised a lighter touch and a focus on auditing area-based outcomes rather than the performance of individual councils. All areas had to report on all indicators, but they were allowed to prioritise some, which were the main indicators included in the Local Area Agreements. However, in terms of QLIs, the same pressures existed, because BVPIs have merely been swapped for LAA indicators, though the process of developing the LAA and the increased emphasis on local collaboration may have pushed some authorities a little further in collaborative practices and this may have created some space for locally-driven indicators, although the reverse is true in the case study outlined in this book.

Local community indicator projects

Community indicator projects have proliferated worldwide, many of them stimulated by LA21. In 1998, MacGillivray estimated that 10 per cent of the UK population were covered by these initiatives in their local areas (whether they knew about them or not). The burgeoning number of community-developed indicator projects is interpreted as the desire of local people to overcome the frustration they feel with national objective/economic indicators that do not match their everyday experience (MacGillivray 1998). They are also seen as a local response to the 'dominance of neo-liberal globalization pursued at the expense of social and ecological sustainability' (Keough 2005: 65). However, indicator projects vary widely, from community-led bottom-up approaches to LSP/expert-led projects with community consultation. Some projects have found it difficult to engage the local community or to retain local commitment. There are many reasons why indicator projects are developed, including to respond to central government agendas. So it is important to look specifically at each case and its local and political context. This also makes it difficult to draw generalisations. However, it is important to be aware of potential problems and issues. These will be discussed below in a review of a well-documented indicators project.

LITMUS

In 1998 the London Borough of Southwark engaged the New Economics Foundation to support a community indicator project called LITMUS (Local Indicators to Monitor Urban Sustainability). The project was jointly funded by the European Commission 'Life Programme' and Southwark Council to explore the use and role of community indicators. A team of four full-time staff were engaged from 1998 to 2000 on an action research project to develop indicators with the involvement of residents in two regeneration areas of Southwark (Peckham and Aylesbury).

Considerable time was devoted at the start to raising awareness of the project through mailshots, local newsletters, stalls and helping to organise community events. Stakeholders were identified and indicators developed through activities that included a survey, focus groups, working in schools, providing facilitation training to local 'community champions' and using 'innovative consultation techniques' such as video and art work and a 'quality of life market'. Participation was invited using a 'quality of life' theme because the perception was that local communities would not respond to the abstract notion of SD. Thirty-six indicators were developed but only sixteen were monitored.

During this initial period (of approximately 10 months) 1,800 residents and 55 organisations participated. However, significant numbers dropped out and it proved hard to engage people on a long-term basis. Only sixteen residents and nine organisations were involved for longer than two months. The project evaluators concluded that there had been a 'modest "shallow" engagement with the community and a very "in-depth" engagement with a small pool of committed people and organisations' (Sommer 2000: 489). More active participation was apparently inhibited by a lack of understanding of the language and process of developing indicators. There was also distrust of the project because it was based in the council and well funded, which caused problems for some, particularly in the voluntary sector. There was insufficient time for the research to assess the effect that the indicators had on policy but it was perceived to be minimal. Many of the indicators failed to meet the initial targets set by the project: that they should be action-focused, have social meaning, be measurable, and be understandable/ simple.

However, a degree of trust was built over time with the remaining participants, and this trust was crucial to the development of working relationships and good local relations. Participants reported significant 'by-products' including increased skills and confidence, improved social networks and trust, and information-sharing. The people involved in developing the indicators had become more active in the community but had not changed their environmental behaviour.

> The impacts generated by the LITMUS process, such as information, networks, trust and skills, have been much more important in influencing the regeneration programmes than the indicators themselves.
>
> (Lingayah and Sommer 2001: 33)

In their research project with 60 local authorities to identify this 'missing link' between indicator development and policy impact, Higginson *et al.* (2003) found that the following factors hampered indicator effectiveness:

- Indicators were not embedded in mainstream policy process, but developed in isolation;
- Quality of life is complex, some issues are outside the remit of the local authority;
- Indicators were not mandatory and therefore low on the agenda and not taken seriously;
- There was often little senior management buy in, and service heads were disengaged;
- There was a lack of resources to develop indicators properly;
- There was a lack of clarity about who was responsible for what;
- There was a lack of vision and leadership;
- The indicators were often not communicated widely or effectively.

The report recommended that the status of QLIs should be raised by linking them into existing structures and policy developments; improving communication to a wide audience and strengthening leadership of indicator projects.

Indicators are developed in the context of wide variations in arrangements for local governance such as: how LSPs work, how effective the community strategy is in joining up policies, how sustainability is incorporated, and the different ways that local authorities respond to the two discourses of sustainability and modernisation. In addition, the relation between indicators and public policy is very complex, spawning a variety of different conceptualisations. For some actors it's a linear relationship – a positivist/objectivist content with a top-down delivery system, privileging national over local imperatives. For others, it's a multi-directional relationship – a socially constructed content emerging from interactivity between stakeholders, privileging local community over national imperatives. Those who see the policy process as inherently complex and dependent upon the political and social context tend not to expect indicators to have a direct influence, no matter how scientifically robust they are. So for them it becomes less relevant to talk of indicator success or failure in terms of concrete effects on policy. A successful indicator:

- Frames a problem so that sufficient consensus can be reached about its 'definition, explanation and solution';
- Appeals to scientific rationalities but is capable of being used dramatically because it reflects social concerns;
- Is institutionalised in an active and effective public policy;
- Remains present in the main public arenas, particularly general media of communication.

Research contextualisation of indicators

There have been many indicator projects where academic researchers have worked with practitioners and local communities to aid in developing indicators (Williams *et al.* 2008; Rey-Vallette *et al.* 2007; Astleithner and Hamedinger 2003; Sommer 2000). My own case study is one such project. However, some writers have identified potential problems with the research/practice relationship, focusing particularly on the perceived gap between research knowledge and influence on policy. Williams *et al.* (2008: 113) discuss the 'traditional *unidirectional* transfer of knowledge' between researchers and practitioners as being an ineffective strategy for influencing policy. The notion of research 'dissemination' relies on a one-way process of sprinkling knowledge on receptive and fertile ground, and largely ignores the complex political context within which policymakers are working. Instead, several authors argue for a two-way mutual transfer of knowledge, whereby practitioners and researchers are collaboratively involved in problem-solving. Research is then context specific, policy relevant and owned and understood by policymakers. Researchers and policymakers alike might assume that participation is something to be done with/on ordinary people and that policymakers simply act on the results of it. But White and Pettit argue that participatory research to develop indicators must also engage with practitioners:

> The generation and integration of appropriate data are not enough: there is also a need to strengthen the engagement and relationships among key actors within processes of research, policy and practice. This means that 'the people' should not be the only participants in the research process. Rather, participatory research ought to involve key officials as stakeholders within the design and process to help them own the findings, and influence knowledge and action at the levels of policy formulation and programme implementation rather than relying on the research report to achieve results...Officials and middle managers are often those who could best benefit from an enhanced understanding of poverty and wellbeing, and from more experiential immersion and knowledge of poor people's realities.
>
> (White and Pettit 2007: 260–261)

Interactive research promoted by the PASTILLE consortium of researchers/ practitioners (2002) is a method created to achieve this engagement between researchers and key officials. Taking a social constructivist approach, it aims to break down the perceived but false division between objective research and practice. The research drew on case studies from local authorities/municipalities in Europe (including one from the UK) and focused particularly on the benefits and challenges of academic/practitioner research alliances whose purpose was to develop indicators. It unearthed tensions between academics and practitioners due to, for example, different 'languages', different institutional pressures, different expectations for the research process and outcomes. It proposed that these tensions are partly to blame for the notorious gap between research and practice and found that these differences can lead to 'an ambivalent relationship' and at times to 'severe conflicts'. Researchers are often kept at a distance and put in a role of 'consultant' in order to provide 'an objective basis for policy decisions', which betrays incomplete appreciation of the role of collaboration in producing effective indicators. Social scientists who are committed to more participatory research are frustrated by this but may not understand the time pressures (due to different institutional expectations) and the pressure for results rather than reflection, faced by officials.

Therefore there is a need for the research/practitioner collaboration to be strengthened for the effectiveness of indicators. The PASTILLE approach identifies three forms of knowledge which must work together for indicator development:

- 'Socially accredited knowledge' – based on expertise around professional/ educational accreditation, for example statistics, demographics, housing, economics, engineering, ecology, air pollution etc.;
- 'Experiential knowledge' – acquired from 'being' and 'doing' based in the locality and local culture. Often described as lay or local knowledge;
- Knowledge about policy processes themselves, knowing how administrations and networks operate and where power lies.

A strategy of 'interfacing policy, research and community' is advocated by Williams *et al.* (2008: 114). They suggest several knowledge-transfer techniques, including the use of 'community policy forums' and the employment of action researchers as 'policy entrepreneurs' who can communicate research findings but also have the power to promote ideas through policy networks. So, the case study outlined here reflects the attempt by academics and practitioners to work together to better reduce this gap, considering the role of local knowledge, policy processes and power as well as research on wellbeing.

Conclusion

The discussion so far has covered aspects of quality of life and wellbeing, and sustainable development. These are contested subjects whose definitions remain elusive despite the considerable efforts of philosophers, politicians and research practitioners over the last fifty years (and earlier). To those charged with the task of developing indicators, this lack of clarity presents a considerable challenge. A wide range of different interests must be brought together in order to reach an 'overlapping consensus', but this process is fraught with issues of representation, knowledge and power. Participatory democracy is essential, but may be undermined by the same central discourses that promote it. This means that in order to develop effective indicators in a local context, it is necessary to examine the predominant discourses and consider how knowledge and power work to privilege or subdue local voices. Previous research has drawn attention to the value of such an approach in advancing the effective use of indicators.

> The discursive approach to sustainability indicators is still in its early stages, but further studies in this direction could help to understand better the implementation of sustainability policies and the relations between power, politics and knowledge in the sustainability case.
>
> (Ortega-Cerdà 2005: 13)

The acceptance of indicators as 'discursive elements' (Ortega-Cerdà, 2005) rather than objective technical tools would increase and strengthen the commitment to the democratic process of their development, rather than the predominant focus on the product by policymakers. However, the review outlined in the first few chapters of this book has highlighted the opportunities and challenges associated with such an endeavour. There are a considerable number of obstacles to be considered and these will be now explored in practice in the case study that forms the rest of the book. Indicator development is a potential struggle and it is this struggle which holds the potential for change:

> Sustainability Indicators do not and cannot in themselves drive policy. Rather it is the process of developing and using them and the way that this process subtly changes the relationship between actors that is the important potential catalyst for change. Hence various aspects of these relationships – such as conflict, mutual (mis)understanding and the search for legitimacy – are the relevant foci for researching and assessing this potential for change.
>
> (PASTILLE 2002: 15)

Part II

Measuring wellbeing in practice

6 Case study of Blyth Valley Borough Council

Background and context

Introduction

In the following chapters I give an account of a research project in an English local authority which aimed to develop a local definition and measurement of wellbeing. It explores and illustrates some of the issues raised previously around understandings of wellbeing and sustainability, participation, knowledge and power in the policy process. The project was a three-year collaboration between the Politics Department of Newcastle University and Blyth Valley Borough Council (BVBC), a small district authority in the Northeast of England. The council wanted to develop a set of indicators to reflect a local concept of wellbeing, within the paradigm of sustainable development, and to do so through a genuinely participatory process. They believed a collaboration with a university would provide extra capacity to explore some of the challenges. Academic staff in the Politics Department developed a funding proposal together with officers at BVBC, which was awarded funding from the Economic and Social Research Council. I joined the project in 2004 as a PhD researcher to work with the council team charged with developing the indicators. This chapter describes the context of that study. Chapter 7 describes the indicator project itself, and the processes and controversies involved. Chapter 8 focuses on the outcomes and impact of the project, and how these related to understandings of wellbeing and sustainability. Chapter 9 explores the difficulties of producing a set of indicators and looks at policymakers' conceptions of indicators and their role in policymaking.

Blyth Valley Borough Council required both practical and theoretical outputs: namely, a set of quality of life indicators, with an evaluation of how effective they would be in contributing not only to sustainability and wellbeing but also to local democracy and policymaking. Its policymakers entered into a difficult contract with the university: to begin a process of critical reflection on their policies and practices, not with the university as an outside 'expert' but with a researcher who was gaining direct experience of the challenges by participating in the work. I worked part-time at the council and joined in the work of developing an indicator project, which included consultation and focus groups with the public. I also carried out in depth interviews with 31 policymakers in the council and wider partnerships, to gain deeper insights into some of the procedural and policy context from differing perspectives.

Over the course of the project the policy landscape changed rapidly, with a plethora of new initiatives issuing from central government and various rounds of restructuring in the council. This altered the original context of the research and continually affected the understanding and expectations of those involved. Governance and organisational structures altered, and new partnerships were brokered while others broke down. Individual relationships and working alliances changed. Nevertheless, some contextual aspects endured including a number of institutional norms and beliefs associated with the political and historical development of the area. Bearing in mind these complexities, I sketch out below the social and political context to the study.

Blyth Valley: a story of decline

Blyth Valley is a small area within the county of Northumberland on the northeast coast of Britain. Northumberland is the least populated of England's counties and one of the largest. It is host to a heritage coastline, important historical sites, and has a richly varied landscape and wildlife; tourism is an important part of the county's economy. Blyth Valley occupies a tiny (70 km^2) south-east corner of Northumberland. With a population of 81,000, it is a comparatively densely populated post-industrial area within a largely rural county. It covers only 1.5 per cent of the area of Northumberland yet holds 26 per cent of the population. Situated on a coastal plain, its main town, Blyth, was once a centre of excellence in shipbuilding, and the first aircraft carrier of the Royal Navy, HMS *Ark Royal*, was built there. Its position within the southeast Northumberland coalfield means that a third of the area's employment was once associated with the coal-mining industry. At its height in the 1960s, Blyth was one of the busiest coal ports in Europe. Like many other UK industrial areas, Blyth saw a relentless decline in prosperity during the late twentieth century. It lost its railway link to the Beeching efficiency cuts,[1] and the gradual (and controversial) closures of shipyards, mines and power stations destroyed many thousands of jobs. The contraction of the mining industry alone accounted for over 10,000 job losses across Northumberland, a figure which represented 32 per cent of all jobs occupied by men (Beatty *et al.* 2007). In addition, the economies, cultures and communities which had grown up around these industries suffered (Waddington *et al.* 2001; Bennet *et al.* 2000). Cinemas, shops and dance halls closed. Important social events, like the annual miners' gatherings, struggled to retain their meaning. Bitterness at the Thatcher government following an unsuccessful year-long struggle of industrial action to stop mine closures in 1984/1985 still endured. The closures left a post-industrial landscape with derelict and contaminated sites, many of which bordered on social housing estates, once lively but now in decline. Although environmental reclamation was carried out between 1970 and 1985 on former colliery sites, this was basic by today's standards; the problem of acidic groundwater leaching was widespread and brownfield sites proved difficult and expensive to develop. By the 1990s, Blyth Valley was suffering severe economic disadvantage, social deprivation and environmental degradation. Its relatively isolated geographical position, lack of amenities, post-industrial landscape and environmental problems meant that it did not benefit from tourism income, and attracting private investment was hard. It was therefore heavily reliant on externally funded regeneration programmes for coalfield area support which, although they have improved the labour market elsewhere, have had only limited success in Northumberland. In 1997, a BBC Panorama documentary highlighted the social issues in Blyth, focusing on widespread drug problems that had caused the deaths of 17 young people over a two-year period (Taylor 2005). Blyth was sadly dubbed 'the drugs capital of the North', a label that has lingered for more than a decade.

The rocky road to recovery

From this low point, Blyth Valley Borough Council and its wider partnership determined to turn the borough round. Behind this ambition lay a multitude of contested views and concepts, some of which I explore in this chapter. When the Local Government Act 2000 came into force with its focus on partnership-working for wellbeing, Blyth Valley had a head-start as it had already worked closely with other agencies to tackle the problems which existed. The newly-formed local strategic partnership drew on many of these established relationships. It aimed to regenerate the economy and increase employment by attracting investment through public-private partnerships. A pro-active strategy on renewable energy, which took advantage of Blyth's coastal location, saw two prominent wind energy developments come to fruition. In 1993, Blyth Harbour Wind Farm, consisting of nine 0.3 mW turbines, was established along the East Pier of the Port of Blyth. In 2000, two 2 mW turbines were installed one kilometre offshore; at the time this was the largest offshore wind energy development in the world. In 2000, a regeneration body, the South East Northumberland and North Tyneside Regeneration Initiative (SENNTRi),[2] was formed to create a 'socio-economic step change' in the wider area, a key aim being to 'improve quality of life'. Its board included representatives from the local authority, Regional Development Agency and English Partnerships.[3] In 2002, SENNTRi published their 'Corridor of Opportunity' prospectus for linking this post-industrial sub-region, which included Blyth, with the city region of Tyne and Wear to generate economic opportunities and investment. Indicators of success included jobs created, amount of private sector investment, number of residential units and area of commercial floor space created. These fed into regional indicators of economic activity such as GVA[4] statistics. Building on these initiatives, for example, the National Renewable Energy Centre was set up in 2002 in Blyth by the regional development agency One North East to train people in energy-saving technologies. The centre has since grown into a comprehensive facility offering a range of services to businesses involved in the renewable energy sector. In addition, various environmental schemes to improve the visual landscape of Blyth Valley such as the 'Greening for Growth' initiative, were implemented in order to attract investment. Blyth Valley Borough Council also endeavoured to 'get closer to communities' through a programme of community development work for which they were awarded 'Excellent' status in the national Beacon Scheme in 2006.[5] Central to this programme were participatory methods including interactive websites and panels of residents who were regularly consulted on particular issues. The borough made significant investments in partnerships in order to boost economic growth, improve environmental quality and empower its community.

Employment and investment are slowly increasing. However, significant problems persist, and some wards in the borough are among the most disadvantaged in Britain,[6] with high rates of unemployment, poor health and children on the child protection register. Extensive work by Beatty *et al.* (2007) on the economic recovery of the coalfield areas showed that despite evidence of a limited recovery in terms of the labour market in some areas, including Northumberland, overall (at the time of the study) none of the job losses associated with the coal-mining industry had been replaced. The problem may in fact be greater due to 'hidden unemployment' including that resulting from the diversion of thousands of people from the unemployment register to incapacity benefit. In Blyth Valley alone, from 1981 to 2003 an estimated 4,100 (of an 81,000 total population) were diverted from unemployment to sickness benefits (Beatty and Fothergill 2005). Of course, there are complicated reasons for this, not least the now well-established causal link between unemployment and ill health.

The 2004 Comprehensive Performance Assessment (described in Chapter 5) rated BVBC as an 'excellent' authority. However it stated that:

> The council is not effectively measuring the impact that its improvements in services and initiatives are having on the quality of life of local people.
>
> (Audit Commission 2004: 19)

The consequent quality of life indicator project described here is evidence of BVBC's desire to understand how such measurements could be created to reflect the impact of its development strategies. Measuring wellbeing in this context meshes with dominant development and regeneration narratives, and also with national performance measurement. This indicator project is thus set within a wider political and institutional context.

Administrative and political uncertainty

Blyth Valley Borough Council was one of six district councils operating under the umbrella of Northumberland County Council (NCC). This two-tier administrative structure of counties and districts was set up across most of England and Wales in 1974.[7] The counties were responsible for services that included education, transport, strategic planning, fire services, social services and libraries, whereas the districts were responsible for others such as local planning, housing, local highways, environmental health and refuse collection. Local government reorganisation between 1995 and 1998 replaced several two-tier authorities with

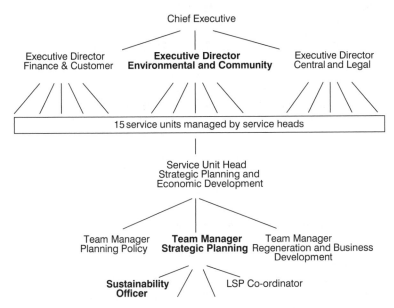

Figure 6.1 The officer structure of BVBC, showing key people involved in the QLI work (in bold)

single-tier unitary authorities. A further round of local government restructuring was proposed in 2007 and had a large impact on this study. As a result of the 2007 Local Government Review, Blyth Valley Borough Council (BVBC) merged with the other district councils within NCC to form a single unitary authority, Northumberland Council.

Blyth Valley Borough Council employed 500 staff, many of whom had been there for many years. The authority was Labour-led by a cabinet of seven councillors. The executive management team consisted of four people: a chief executive and three executive directors with responsibility for finance, legal and service provision respectively. Below this management tier were 15 service heads, each managing a service unit. The work of developing the indicators was located in the Strategic Planning and Economic Development (SPED) service unit. The duties of this unit included supporting the LSP, developing the (Sustainable) Community Strategy, ensuring the integration of regeneration strategies and plans, and providing research and data analysis to inform strategy development. It was also responsible for sustainable development and climate change strategy. Three council officers were included in the team along with myself to develop the indicator project: the executive director with responsibility for environmental and community services, the strategic planning manager, and the sustainability officer (see Figure 6.1). The indicator project was to be reported on regularly at the LSP meetings and it was initially envisaged that, alongside wider community participation, the LSP would be collaboratively involved in its development.

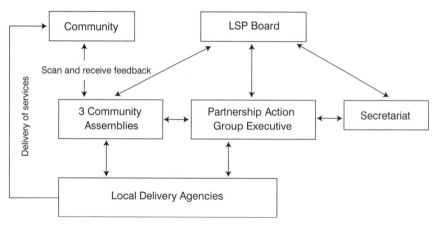

Figure 6.2 The structure of Blyth Valley LSP 2004

Blyth Valley local strategic partnership

As described in previous chapters, the Local Government Act 2000 required every local authority to set up a local strategic partnership and to develop a local community strategy following wide consultation about local quality of life. Both LSPs and community strategies have been described as having varying degrees of success and workability from area to area (Geddes *et al.* 2007; Darlow *et al.* 2007). As these were designed as the main mechanism for discussions about strategic wellbeing policies in the borough, I now go on to consider how they worked at Blyth Valley. Figure 6.2 is a graphic produced by BVBC in 2004, when the indicator project started, showing how the LSP was intended to work.

The LSP Board were responsible for strategic policymaking and met quarterly. It was ultimately accountable for the delivery of the Blyth Valley community strategy, named *The People's Plan*. The chairperson was the elected leader of BVBC; the other seven voting board members of the LSP Board represented key sectors (business/private, community/voluntary, local authority, police, education/ skills, health and social services). The Priority Action Group Executive (PAGE) represented a wider section of the professional community in the public, private and voluntary sectors and was responsible for creating action plans to ensure that *The People's Plan* was delivered. These action plans would then be fed down into the services by the local delivery agencies. There were three community assemblies, one each for the three geographical areas of BVBC, (Blyth Valley, Seaton Valley and Cramlington). Community assembly meetings were convened quarterly by elected members of the council. Anybody in the area could go along and air their views. The community assemblies were intended to be 'the eyes and ears' of the LSP in the local area and to provide for community representation. The secretariat, which included an LSP co-ordinator and an information officer, worked behind the scenes to support the partnership. On paper this seems a reasonable structure which answers the need for a strong strategic steer that is closely informed by local needs, backed up by debate, linked to action, and serviced by adequate resources. But in order to be effective, such a structure must also be supported by a culture of political engagement and trust in political process.

There were major problems with the functioning of the LSP. The board members often did not turn up for meetings and most of those that I attended over a three-year period were inquorate. Several people described this as a 'capacity issue'. One member of the LSP explained that the two-tier structure of local governance in Northumberland meant that, as he represented a regional organisation, he was expected to attend three LSPs and that there were not enough staff in his team to become as fully involved as they should with every district authority area. There was dissatisfaction among the secretariat officers with the situation; they were constantly trying new ways to engage board members and encourage them to form strategic policy. The LSP chair (and leader of the council) rarely attended Board meetings. The LSP lacked the capacity or political backing to make a strategic local input and essentially functioned as a source of information for members and as an agency for rubber-stamping various policies.

> I think most of the policies and plans and strategies are about the local authority delivering a target not about what needs to be delivered, or the LSP saying, 'yes we now have set this up', tick box.
>
> (LSP member)

The PAGE group met more regularly and had already undertaken work in developing action plans. However, it was unclear how these plans would be implemented. A common question at PAGE meetings was what 'added value' the LSP was bringing. There was a sense that the smaller partnerships which had been formed to address specific issues such as primary health care and crime reduction, worked effectively. But it seemed hard to link action on specific issues to a strategic overview. Some officers felt that there was little joined-up thinking and that real strategic action was rare.

Meanwhile, there was widespread cynicism about the community assemblies, based on a perception that they did not represent community views sufficiently and neither were they truly empowering communities to become engaged in local governance. Frequent comments like 'all the decisions have already been made,' and 'they are just talking-shops' illustrated perceptions that the community assemblies were dominated by a few residents and local councillors 'who really are representing themselves and not the whole community' (BVBC officer) and did not promote participatory democracy. The local authority scrutiny team were undertaking a substantial review of the assemblies during the indicator project. At the start I was told clearly by several people that these were not the right fora in which to develop discussions about the indicator work. In addition, the LSP and wider democratic arrangements were undergoing a great deal of uncertainty. Some of that political context is described below.

The Northeast Referendum and Local Government Review

In November 2004, shortly after the beginning of the indicator study, as part of central government reforms to devolve power to regions, the Northeast held a referendum to decide whether or not it should have an elected regional assembly. The Northeast was the first region New Labour selected for this experiment. It was chosen partly because it has a distinct regional identity and was perceived as a Labour stronghold.[8] Despite expectations, central government's proposal was resoundingly rejected by 78 per cent of the vote. This was a major blow to New Labour's plans for devolution throughout other regions, and these were subsequently scrapped. Ironically, subsequent research shows that one of the main reasons why the region rejected a regional government was because residents were disaffected by politics in general and certain national politicians in particular. New Labour's attempt to promote democratic renewal was defeated by the very democratic deficit they were trying to correct (Shaw and Robinson 2007; Tickell *et al.* 2005). The result also highlights the apparent gap between political elites in the north (who were in favour of the proposal and whom the government had canvassed before deciding on a referendum) and ordinary residents. The latter were unsure what an elected assembly might offer the region and were as distrustful of local politicians as they were of central government (Tickell *et al.* 2005; Rallings and Thrasher 2006). This left the existing regional institutions in some disarray. An unelected regional forum had already existed, which had been renamed 'assembly' in anticipation of a favourable vote. After the referendum result, the legitimacy of this body which had no democratic mandate, came into question. Subsequently there was no effective, coherent forum for regional governance (Shaw and Robinson 2007). As part of the referendum, people in Northumberland were asked to vote on whether they wanted a single tier of local government. Again they voted no. The post-referendum climate affected the local authority in a number of ways.

The two-tier system of local government was highly unpopular with officers at BVBC and relationships between the county and district councils were at times tense. Many felt that the social needs and aspirations of the post-industrial Labour-controlled Blyth Valley were different from those of the largely rural, Conservative and Liberal Democrat county. Most of the officers in BVBC wanted to abolish the two-tier structure and have more control over policy. As one said, 'It's a disaster for us.' Blyth Valley councillors and officers felt that NCC was not in touch with the needs of people in the former industrial areas of Blyth Valley. Officers in BVBC wanted to associate Blyth Valley more closely to the Tyne and Wear city region than to rural Northumberland. BVBC's district neighbour, Wansbeck District Council, has a similar geographical, historical, and socio-economic profile. The two councils had worked together closely on particular issues and already had joint initiatives to promote primary health care and reduce crime. At the time of the referendum they were considering combining their LSPs into a single one to represent southeast Northumberland. According to the chief executive of Blyth Valley they shared 'very similar issues in terms of social and cultural history, aspirations and quality of life. My personal view would be not to try to kick start our LSP; it would be about…developing a real strong strategic function in southeast Northumberland'. This partnership was formalised in the lead up to the 2007 Local Government Review conducted by the Department of Communities and Local Government. The district councils submitted a proposal to central government for the creation of two new unitary authorities, South East Northumberland (Blyth Valley BC and Wansbeck DC combined); and Rural Northumberland (Tynedale DC, Berwick BC, Alnwick DC and Castle Morpeth BC). Northumberland County Council submitted an alternative proposal for a single-tier council for the whole of Northumberland. This was a time of increased tension between districts and county, and great transition and uncertainty for officers and councillors. This impacted heavily upon the study described here as well as on morale within the council and LSP:

> I do think [in] the last couple of years though, that the impetus surrounding, say, our LSP, the whole of the community leadership stuff, has been derailed a little bit…And whether that's because of…the referendum, its effect…but local government in Northumberland and the pressure and tension that that placed on partnership relations, especially County and District…And now we've got another White Paper coming along in June which can be about the future of local government; really that sort of issue just puts paid, I think, at times to real innovation in terms of partnership working.
>
> (Senior management team member)

> I think a lot of people in Blyth Valley have an idea that they control, they can control politics, which is the way it should be…but I think if things went distant, if we lost the locality of the council and the area, I think that would have an untoward effect, a real bad one, on the people down here.
>
> (Cabinet member)

In July 2007, the Department for Communities and Local Government decided to opt for the proposal submitted by Northumberland County Council, and create one single authority for the whole of Northumberland. This caused much resentment, particularly as the referendum vote had been against a unitary authority. With an area of 5030 km^2 Northumberland is now the largest unitary authority in the country, making its claim to be a 'local' authority spurious; it is also the 'victim of the harshest democratic cull' as all the district councillors have gone and the number of county councillors has not increased (Game 2009: 21). There has been a shedding of 551 (89 per cent) councillors, leaving only 67 to represent 307,000 people. This means that instead of one councillor per 496 people there is now one per 4,580 people. This is in comparison with a UK average of 1:2900, 1:415 in Germany, and 1:117 in France (Game 2009). In addition the political climate in Blyth Valley has radically shifted. Whereas it had been previously governed by a Labour-controlled council, it became a labour minority within a Liberal Democrat/Conservative county, a bitter pill for those still harbouring resentment for the Conservative government that closed the mines.[9]

In April 2009, Blyth Valley Borough Council ceased to exist and many of those involved in the indicator project either lost their jobs or were redeployed. This case study is therefore set in the last few years of Blyth Valley Borough Council, at a time of great political and administrative upheaval and uncertainty. Discourses of how Blyth Valley was going to develop in the future were interconnected with these political–administrative struggles as many of the senior officers and councillors at Blyth Valley wanted to align Blyth Valley to Tyneside rather than rural Northumberland. I now outline the dominant development discourses in the council, illustrated by a high-profile regeneration project which was happening at the same time as the indicator project.

Twenty-first century sustainable communities: regeneration policies

Working closely with Blyth Valley Borough Council (BVBC) and the regional development agency, the Blyth-based sub-regional development organisation SENNTRi commissioned a study to look at regeneration potential along Blyth estuary, an area including large amounts of ex-industrial land. The Blyth Estuary Development Framework Plan, published in January 2005, outlined 'a range of opportunities available to unlock the potential of the sub-region and transform the area into a successful, progressive and sustainable community with a strong local identity' (Llewellyn Davies 2005). Bates Colliery, the site of a derelict and contaminated deep coal mine next to the river, was identified as a potential residential and retail development 'flagship' for this transformation. The site was described as 'an opportunity to create a regenerated sustainable community for the 21st century as a pivotal component of the wider renaissance of South East Northumberland and North Tyneside region' (SENNTRi 2006). During the indicator study, a controversial project was being proposed as part of this development along Blyth Estuary, involving the relocation of 295 households and the demolition of their homes. I followed this project and the wider regeneration plans closely, attending steering group and planning meetings and resident forums. I was interested in how discourses of quality of life and wellbeing were mobilised in these proposals and in the ensuing controversy. I have written about this in detail elsewhere (see Scott 2009 and Scott 2012) but I will summarise some key observations here to show how the institutional context determines whether or not a set of indicators can be influential, and how, in order to work, indicators must appeal to particular discourses.

Although the institutional discursive arena was complex, involving a diverse range of actors and influences from the local to international level, policy discussions were dominated by an economic growth discourse that focused on place-making and physical regeneration. This was rooted not only in 'Old Labour' views of joblessness as a structural issue but also in New Labour notions of active communities. Objective socio-economic indicators of deprivation were often used to justify the need for radical transformation. The focus was on attracting private investment and affluent people into the borough in order to create 'mixed communities' which would not only raise economic growth indicators but also, crucially, change a culture of low aspiration. Inherent in these narratives was an under-developed notion of quality of life, based on reducing socio-economic deprivation. They were heavily influenced by New Labour discourses of community responsibility, social capital and participation. Ironically, when questioned about the finer detail of how the wellbeing needs of the disadvantaged would be met, or inequalities reduced, policymakers could make no more than tenuous causal links based on the idea of a 'trickle down' of greater affluence to the poor. There was a general acknowledgement that economic growth did not, in itself, address inequality issues and that other interventions (usually meaning community development work or neighbourhood management schemes) needed to be in place. Policymakers in SENNTRi and BVBC seemed unsure how large-scale physical regeneration would impact positively on the most disadvantaged economically in the borough, rather than simply making the wards look better in terms of the objective indicators. There was an acknowledgement that they were unconvinced by discourses of economic growth through place-making as a means to address individual deprivation; at best they didn't really know ('there is a huge lack of evidence') and at worst this was a fallacy:

> I will accept that the concept of trickle-down does not really work because what's happened nationally is that economic success has tended to result in either two incomes or no incomes.
>
> (Senior policy actor)

While it was obvious that regeneration plans could bring important benefits, and that there was existing deprivation which demanded action, the potential negative effects of the plans were under-represented and under-discussed. In this case, the rationale for the demolition came from discourses about quality of life and was justified by evidence of the high levels of deprivation and infirmity on the estate, yet ironically little attention was paid to the impact of the planning and implementation process on the wellbeing of these same residents:

> It got to the point, me and my wife a while back got very close to splitting up...a lot of it was depression caused by this pushing it over the top. This regeneration thing has caused really big problems particularly with her, basically, she ended up on tablets for depression because of it. A lot of people might say 'oh that's a bit drastic' but everybody's different. Like I say we're not particularly big, ambitious people, we started off getting married, she was sixteen, I was eighteen, we had nothing, we lived in a little flat, luckily we got a house in Hodgsons Road, lovely house, lovely area, lovely neighbours, eventually we bought it and then I came into some money and we decided to do it up, not to sell it, but to improve the quality of life, nice little conservatory, sit in the sun, redo the kitchen, got the plans done, paid the deposit for the conservatory, then this letter lands on the doorstep...It has caused a lot of problems, a lot of problems.
>
> (Local resident)

In addition to the impacts of the regeneration plans on the residents themselves, some policymakers, while wanting to get rid of 'poverty ghettos', were also concerned that if deprivation was scattered throughout the borough then it would become less visible. This makes ward-based IMD statistics look better but it becomes harder to argue for nationally distributed intervention funding. Paradoxically, it is not known whether funding targeted towards the most deprived areas (bringing the pressure to meet particular targets) combined with a 'mixed communities' policy, merely disperses rather than solves deprivation (for a review of evidence see Tunstall and Lupton 2010; Cheshire 2007).

The wider regeneration project also typically produced environmental controversies because the site was environmentally sensitive in a number of ways. This caused concern to local environmental stakeholders and some residents. The area is vulnerable to rising sea levels and contains retention ponds of polluted water pumped up from old mine workings. Pollutants from active industrial workings nearby were causing health worries to existing residents, and many questioned how this could be reconciled with plans for new residential development. There were also concerns about wildlife. Skylarks had been nesting on the land for the previous few years, and the estuary includes a Site of Special Scientific Interest (SSSI)[10] that provides mudflats for winter migrant birds. Environmental organisations, such as Groundwork Trust and Northumberland Wildlife Trust, and environmental representatives from the district and county councils, had been working on an estuary master-plan to create a strategy for protecting biodiversity, dealing with flood risk and increasing eco-tourism to the area through a heritage trail. Many, including the sustainability officer for the council, argued that the site should be ecologically restored and become a tourist and wildlife amenity:

> I just think if they wanted to do something radical…that it would be the most amazing thing to have this whole area as some kind of estuary park and people from Blyth Valley, that northern edge of Blyth with the least money, would benefit the most… I would see that as an added value for anybody if you were trying to entice people into Blyth, to actually live and work in Blyth, it would be something good for you to stay here.
>
> (Sustainability officer)

> They [policy actors] should think outside the box and use the environment as a regeneration tool…they should do what they know is right and good things will come on the back of it.
>
> (BVBC officer)

However, many policy actors in Blyth Valley implicitly and explicitly evoked a hierarchical theory of needs, suggesting that concern about global environmental issues was limited to the more educated, affluent people whose basic needs had been met. In terms of policies, economic development was prioritised over environmental protection because the latter was seen as less relevant in addressing the basic needs of people in Blyth Valley, predominantly through income:

Whether I've got it right or wrong, not everybody wants to live in an absolutely environmentally friendly way yet, but what they do need to survive in this society is some money in their pockets. And why? Because…you want to make choices and you want to move on. So I come back to that all the time.

(Executive director, BVBC)

[The environment is] one of many priorities but it's not the highest, the highest is keeping people safe and warm in the area, providing them with employment, keeping the economy going, that to my mind, I'll be honest, is my priority and…then when we've got these things right then we can look more vigorously at the environment.

(Cabinet member)

At the moment we just want to get jobs in the borough, the green stuff is for later.

(BVBC officer)

When you are talking about the people of Blyth Valley for them the problem is not global warming, it's not the hole in the ozone layer, it's a lot of people living from one day to another. And it's trying to set people's horizons a little bit so they can see beyond the end of the year. But it really is down to basics and can they afford to eat this week.

(Councillor)

Despite detailed feedback from the community that told policymakers there were a variety of interests, preferences and needs, when policymakers were arguing for economic development to trump environmental concerns, they used a particular conception of a disadvantaged community to support their position. This conception was of a severely disadvantaged group of people desperate to escape the poverty of their lives and accept new homes at any cost. This idea of a homogenous disadvantaged community became a truth with enormous normative power. Not many environmentalists were prepared to argue against such talk in a meeting for fear they would be branded as 'putting the environment before people', as several officers and members interpreted it. This truth was partly constructed by presenting comparative statistics that showed levels of poverty, education and health in Blyth Valley were considerably bleaker than the UK national average. It did not involve presenting responses from community consultations, and these suggested that the picture was much more complex:

I think with the new building and everything, people might want to tidy up the riverside but the mudflats should be kept, they need to be kept there for the birds.

Anything eco would be great; plenty of green areas for walking, cycling, relaxing.

There were a lot of good ideas. I specially liked the ones about saving rainwater and using green materials.

Therefore, in promoting 'objective' evidence about deprivation policymakers were able to legitimise a particularly simplistic conception of the community, its needs, associated problems and solutions. This was partly due to the promotion of a positive and dominant transformation discourse where the focus was on physical regeneration, culture change and place-making underpinned by a partly formed idea of quality of life based on reducing deprivation:

So of the whole ward we can now pinpoint down to houses and streets that are drawn around, which bit is the most deprived, on multiple deprivation not just on one thing. But you can actually see now where the pockets are…So what I suggested to [colleagues] was I found out where all the super output areas were, and we put together a plan for 10 years and said, 'this is where we're going' and we're going there based on these figures. And we go in there for 6 months, and we make a difference.

(BVBC officer)

The focus on deprivation identified through national statistical indices created a geographical focus on disadvantaged areas, where resources and community development initiatives could be targeted. This work often focused on improvement of the local area and while I was at the council, I followed two such interventions. These were focused on litter clearance, play equipment installation and improvements to community facilities. Community development workers helped residents organise into groups and access training, facilitated links with support agencies, organised community trips out and craft groups. While these interventions were obviously valuable and valued by some residents, there was really a limit to how much change they could affect in an area. The community development workers felt that the most deprived people would often not get involved. Low self-esteem and lack of confidence were cited as main reasons.

Gender, environment and economic growth

We are driven by an economic agenda and the economic agenda tends to be driven by blokes.

(Council officer)

Development conflicts and underlying institutional economic growth discourses of the type described above are not of course unique to Blyth Valley. However, what is specifically interesting in this situation is the way the conflicts were discussed, the additional issues that were woven in, the characters and relationships of the actors involved and the context in which they were set. The indicator project was subject to two distinct sources of tension within the council. One was between environmental interests and the dominant policy discourses of socio-economic regeneration, as already discussed. The other arose from the perception that there was a well-established, homogenous and exclusively male core group of decision makers. This was linked to wider concerns about the nature of democratic representation, community consultation, transparency and equality. The CPA had criticised Blyth Valley for their lack of engagement with the equality agenda and the council had been served a notice by the Commission for Racial Equality to engage with a racial equality scheme of work. While I was there the gender equality issue was also being addressed. In 2006, BVBC officers and members held an initial meeting, attended by the chief executive, to consider gender equality issues in both employment and service delivery.[11] BVBC had one of the most inequitable gender balances in the country. While women made up 69 per cent of the workforce, they only occupied 9 per cent of the managerial positions, with just one woman among nineteen people in the top three tiers of management at BVBC. In addition, the cabinet and the LSP board were all male.[12]

Gender recurred as a major theme in my first few interviews with women professionals, although none of my questions related to it. In an interview with a female councillor, she stated that the council did not 'care about the environment'. When I asked her how this perceived deficiency could be addressed, she said simply, 'more women in senior management'. She believed that women had stronger environmental values because 'they think about the future more'. She gave several examples of this, including a disagreement over an air quality issue where a group of male policy actors had ignored a female environmental health officer's recommendation to refuse planning permission for a development. Another female councillor felt that Blyth Valley had a 'very male economic regeneration agenda'. Although only a minority made such explicit links between gender and particular policy priorities, the way that gender was implicated in debates about the environment was re-enforced by the fact that the environmental protection team head and the sustainability officer were women and both quite outspoken. Many people I spoke to felt that having more women in senior positions would increase opportunities for a range of different values, including environmental ones, to be more effectively considered. A number of people felt that the management team was 'inward-looking', 'out of touch', and focused too heavily on economic regeneration at the expense of other issues, including environmental concerns and building the capacity for wider partnership working.

The members of the corporate management team had all worked for the local authority for a long time, had known each other for many years and came from similar backgrounds in the Northeast. In their interviews with me they all commented that they worked well as a team, had complementary skills, and that their working-class backgrounds gave them strong and similar values which informed their work. This positive self-assessment contrasted with others' perceptions that they operated 'behind closed doors' as a 'coalition', 'a gentlemen's club' and 'close friendship group':

I find it really, really hard to break into those networks; they are quite close friendship groups and they don't open out...A lot of them have been around for a long time and that core group who know each other very well are not outward looking, they don't look at a new member coming in and think there's needs there.

(LSP member)

That is something that came out at the performance management thing last week, where there were a few snide comments that nobody knows what goes on in corporate management team [meetings]; the service heads don't get minutes, they don't get told what happens, it's this closed shop, the chief exec, the leader, the deputy leader, the two executive directors cooking up something and then suddenly it happens.

(BVBC officer)

The role of the sustainability officer

The use of the term 'sustainable development' was problematic in BVBC. When I first arrived it was a term that was not in common usage in the council offices or in strategic meetings. In the community strategy, *The People's Plan*, the term occurred only once and appeared self-consciously in inverted commas. When people did use it in meetings, they seemed uncomfortable and often moderated it by saying 'whatever that means'. In meetings when the sustainability officer was present, people mentioned environmental issues more frequently and tended to address only her instead of the whole group. Some male members of staff who worked in the same office often teased her about environmental issues or referred to her as 'the eco-warrior'.

The sustainability officer was responsible for sustainability issues within the council but felt she had 'no particular brief' and that it was up to her to see what influence she could exert. Because the social and economic needs of the area were already high on the agenda and well served by other more knowledgeable professionals in those fields, she saw her role as highlighting issues that were under-represented, particularly the environment. Although she did not associate the term sustainable development only with environmental issues, she was perceived as doing so. She felt that within the council there was a lack of strategic consideration of sustainability. Ironically, policy actors outside BVBC perceived the council as forward thinking in terms of the environment as the sustainability officer was well networked in the area and perceived as pro-active. This gap between outside perceptions of the council and her own, often left her feeling isolated and despondent.

...in some ways I enjoy the strategic stuff and I can see there's a need for it but I'm not going to offer myself up for it, because I know I'd be out on my own and there'd be very little support and it's a hard place to be.

All the LSP voting members were men and although the sustainability officer had never felt that she had been the subject of overt sexual discrimination, she did feel that gender politics presented a significant challenge to her work within the council. She felt that the promotion of a strategic sustainability agenda rested mainly on her, but that she was marginalised because an older male group perceived her not only as an environmentalist but also as a 'young lass' (she was in her early thirties).

Just after the indicator project started the sustainability officer was moved from the Strategic Planning team to the Community Regeneration and Culture Service unit to manage a community-based nature conservation project, although this was not her area of expertise. Having to be responsible for the implementation of various practical community projects made it difficult for her to contribute to strategy. Her absence from the team allowed people to overlook sustainable development issues. For example, in an 'away' event for the SPED team each group manager gave a presentation outlining their current work and priorities. In all, there were six presentations, however, no one mentioned sustainability or referred to their work in relation to any sustainable development framework. This included the head strategic planner who was working at the time on the new Local Development Framework which had a new focus on sustainability appraisal. Afterwards I commented on this to a colleague who had also been present. She thought about it, agreed with my observations and was quite taken aback. She said 'That's really bad, that wouldn't have happened if [the sustainability officer] had been here.'

Other officers felt that the sustainability officer had a difficult role:

I mean she's been given the entire responsibility for the climate change agenda, the sustainability agenda, she has side roles of CONE and various other bits and I understand (the executive officer) has told her to basically bring the whole quality of life objectives on board. One person can't do this. It's ridiculous, we're back to the importance; people don't just add a chunk of work onto your job if they think it's truly important. If they sat down and thought…could she possibly do all this, of course she can't.

(BVBC officer)

I don't think there's an overt campaign to be unsustainable – I think [pause] there's probably a misunderstanding of the word…[the sustainability officer] has been isolated off into 'well you go and clear some woodland and educate a few children' and she's become like the tick box environment box on the Cabinet reports.

(BVBC officer)

I think we like to show that we've looked at sustainable options, we like to show that we've talked to people, we like to put a tick in a box but when push comes to shove, we want the development, we don't care whether we get the sustainable bit. That's where we're at at the moment and that's what's happening.

(BVBC officer)

Summary

The above is brief account from my own perspective, in which I have aimed to show that the project of developing measures to define and evaluate wellbeing in the Blyth Valley did not start on a blank canvas, nor was it isolated from its institutional and social context. Not only was the concept of wellbeing highly contested and subject to a range of agendas, but so were those of participation, community, and sustainability, as was the means of measuring them. The project was also conditioned by the particular ways in which individual actors were represented and involved within institutional discourses. In the following chapter I present a case study of the indicator project itself.

7 Defining 'local' wellbeing

Discourse and debate

When I arrived at Blyth Valley in September 2004 the sustainability officer was disheartened. For the previous year she had been trying to get the Blyth Valley partnership excited about indicators. This would have been a hard enough job without the tensions and uncertainties described in the last chapter. My own experience is that whenever you mention the word 'indicators', eyes glaze over. One friend recommended that I avoid using the word in the title of this book as it would be the 'kiss of death for book sales'.

The sustainability officer had the responsibility for producing a set of 'high-level long-term' indicators to reflect the core goals of Blyth Valley's community strategy, *The People's Plan* (see Box 7.1). After struggling to engage the partnership in discussions about which indicators should reflect this strategy and measure its success, in August 2004 she presented the LSP Board with a set of draft indicators grouped under the five themes of *The People's Plan*, and urged the LSP to 'undertake some development work to understand the value of such indicators and sustainability in general, and determine the right set of indicators for Blyth Valley'.[1]

There had been little interest in these indicators. The sustainability officer believed that the problem was partly due to how her role and sustainability itself was perceived within the partnership. She felt that, as an academic and 'an independent person', I would be taken more seriously. It soon became clear to me that the sustainability officer was keen to use the indicator project (and me) to promote greater awareness within the partnership of the importance of environmental issues and sustainability in general. She talked about this in terms of compulsory education, using phrases such as 'hammer it home', being 'hard-hitting', and 'really jolt people into some sort of awareness'. She understood the critical element of the research, having previously worked as a university researcher and engaged in reflective discussions with me. I found it easy to get on with her and initially we worked closely together. My task was to help this process of indicator development but not only that, to inject new energy into a discussion which had begun to flag. The policy context, as we both understood it, related to the Local Government Act 2000 which, as already discussed, was an attempt to mainstream SD at the local level through a community strategy, so we took it for granted that the project was set within this framework. We assumed

Box 7.1 Five core objectives of *The People's Plan*

In 2003, in accordance with the Local Government Act 2000, Blyth Valley had produced The People's Plan, which comprised five core objectives:

• Establish a high-quality town and country environment that everyone enjoys and takes a pride in.
• Create and sustain local employment opportunities that everyone has ready access to.
• Ensure everyone has the physical and mental wellbeing to tackle the rigours of modern life.
• Afford everyone the opportunity to realise their potential through learning.
• Forge vibrant, distinctive and self-sustaining communities that everyone feels part of and safe within.

that indicators should reflect the local vision of *The People's Plan* while also fulfilling the requirements of sustainable development; in fact this was one of the main tensions that the research project was designed to explore. The sustainability officer was keen that we should not merely produce long lists of indicators as this approach had previously failed to engage the LSP. Instead, she favoured an approach that would initially debate the question of how to define quality of life, then form indicators through a participatory process in collaboration with strategic policymakers. I agreed.

The strategic planning manager, who was also our line manager, seemed to have a more fluid sense of what the project was for. With a BA in politics and experience elsewhere as a local councillor, he readily engaged in philosophical and political discussions with me about quality of life. Like myself, he was on a steep learning curve as he was relatively new both to the post and to Blyth Valley. He was interested in the philosophical issues that the study was raising about quality of life, wellbeing and sustainable development; at the same time he seemed to hope that research would provide a solution to what I believe are essentially political issues, and often expressed this in a tongue-in-cheek way:

> I think every...organisation [whose] role is to improve society as a whole needs...a set of quality of life measures to say that society can be seen to be improving. But I guess you have to identify what you mean by society and what you mean by quality of life and what you mean by improving. Um, hence getting you in, Karen, to answer some of these questions!

He considered collecting 'baseline data' to be a priority. Although an information officer attached to the LSP was working on this, the strategic planning officer would regularly send me indicators and statistics. I was uncertain and concerned about how he saw my role:

Karen's role is to assimilate all inputs, clarify the product and test softer quality of life measures against harder performance measures.

He saw one of my functions as being to facilitate debate and discussion, and initially urged me to go into the LSP Board meetings and deliver a 'provocative' presentation as a way of 'stirring up discussion'.

At the first LSP Board meeting I attended, I was given a half hour slot to introduce myself and the project, and to invite discussion. As an ice breaker I asked each person to briefly tell the rest of us about one of their personal aspirations. There was much engaged discussion and good humour. Many responses related to ambitions to travel. These examples then helped to facilitate discussion about the trade-offs between quality of life and sustainable development, and the difficulties in measuring both. Participants expressed a range of opinions; some felt that Blyth Valley should pay more attention to environmental and fair trade issues, whereas others felt that it should concentrate on dealing with local disadvantage. As one person said, 'let's get our own problems sorted out first, then we can save the world'. In contrast to central government's win-win discourse, these approaches were regarded as different and separate ones for the council, which had limited resources and had to make moral and political choices about which approach would best represent the wishes of its members and the local population.

It was agreed that the indicator project would be on the agenda of every LSP meeting. The sustainability officer and I decided to start the collaborative process by interviewing all the board members about their views on quality of life and what they believed indicators should reflect. We interviewed all voting members of the LSP (apart from the Chair who did not respond to requests for a meeting), and registered a wide range of personal and professional perspectives. We collated all the comments about what board members saw as the main quality of life issues facing Blyth Valley over the next 10–15 years and organised them into themes and issues. We identified some issues that had not been mentioned such as domestic violence, mental health, biodiversity and climate change and wanted to take this comparison to the LSP Board and PAGE, as a way to generate further discussion about what indicators could reflect and who should be involved in creating them. The project seemed to be going smoothly at this point and the sustainability officer and I had a clear idea of what we were doing and how indicators related to the sustainable communities agenda.

Difficult relations

At the same time Northumberland County Council (NCC) was also developing a set of quality of life indicators. As mentioned previously, an impending local government review meant that BVBC was going to either join forces with neighbouring Wansbeck District Council or become part of a Northumberland-wide unitary authority. BVBC preferred the former but there was a sense in which officers, particularly senior managers, were hedging their bets. The project, therefore, became caught up with problematic district/county relations; there seemed to be pressure to compete with NCC to be first to produce indicators, and to be more innovative. This conflicted with the pressure to work with NCC to ensure that the quality-of-life indicator work was integrated. In November 2004, just after I arrived, NCC consulted all the district councils about the set of QLIs it was producing and I had several meetings with its policymakers and researchers. The team of researchers and statisticians that was dedicated to the work on indicators for the county described their approach as 'pragmatic' and were disparaging of 'policy strategy people who dream up indicators which they like but which are impossible to measure' (NCC officer). Their expertise in quantitative statistics strongly informed discussions about how indicators should be created. Their approach was driven by what could be 'robustly measured', and they argued that county and district should be using 'exactly the same indicators'. This did not go down well with policy actors in BVBC, who were keen to highlight the very different issues that South East Northumberland faced compared with the rest of the county in order to justify having a separate authority when it came to the review. I felt that the county statisticians viewed the Blyth Valley indicator research project as not only unnecessary but, due to its qualitative nature, as peripheral to the real, valid and professional work of developing QLI statistical instruments.

Shifting ground

In early 2005 things began to go awry for the indicator project. The executive director, who had attended a number of presentations, became keen that we should make a distinction between the indicator project and the sustainable development agenda. He saw the purpose of the project as being to measure subjective wellbeing, and described this as 'mind contentment' or 'happiness', something that was related to, but distinct from, SD:

> Sustainable development arises from something that we talk about, the planet, it derives from that level of thinking…and I think quality of life is more immediate to the individual. That [sustainable development] is a societal must-do; quality of life is more like an individual must-have. So I distinguish that way…My worry is this idea of individuality versus the societal sort of viewpoint…I think it is going to be a balance between the two but what might be useful is to let others worry about the sustainable development stuff.

In addition, he saw the study as being about developing indicators with the community 'out there' and reporting back to the partnership, rather than producing them collaboratively within the partnership. I felt he saw my research work as being separate from the council's 'pragmatic' need to produce QLIs as an 'output'. He decided that a set of QLIs should be developed from the interviews and presented at the next LSP Board meeting in April 2005. The sustainability officer and I both argued that long lists of indicators had not engaged the LSP previously; we felt it was too early to produce a set of indicators and wanted to continue the process of debate in order to produce a concept of wellbeing that could inform indicator choice. However, the executive officer urged us to 'come off the fence' and present a product that the LSP could 'sign off'; he felt that people were tired of 'talking shops'. The strategic planning manager proposed a middle course, saying that the indicators at this stage were only 'Aunt Sallys' for discussion. But the sustainability officer and I felt obliged to produce a set of indicators quickly, based on the discussion so far. This was an expedient exercise, and neither of us felt confident with the results. We considered all issues raised by the interviewees and scoured existing indicator sets to find the closest proxies for specific issues raised, although many of these indicators were unsatisfactory. As we did not want to prioritise or discard issues (we felt this was a task for political strategists) we presented a long list of 75 indicators grouped under 15 themes to the next LSP Board meeting in April 2005. Despite signing off the indicators as an action achieved, one Board member voiced a concern that these had been produced by 'relying on only seven men's views'. As we anticipated, none of the Board members commented on the content of the indicators, and they were never referred to again.

That month, NCC produced its initial set of quality of life indicators attached to a 'State of Northumberland Report'. As can be seen from the extract below their indicators were modelled closely on the set issued by the Audit Commission and were driven by what was already measurable and legitimised by central government:

> The aim is to provide a set of cross-cutting indicators which combine economic, social and environmental issues…for this sub-regional exercise a set of 27 indicators have been selected based on Quality of Life Indicators from the Audit Commission, DEFRA and the ODPM. The indicators chosen are designed to give an overall 'basket' of indicators which will highlight the quality of life in Northumberland and cover the full spectrum of economic, social, and environmental wellbeing. The aim is to also provide indicators which relate to the Government's headline sustainable development indicators and thereby facilitate regional and national comparison.
>
> (Northumberland InfoNet 2005: 4)

The executive officer expressed his concern that County had 'beaten' BVBC to publishing a set of QLIs and this sense of pressure may have explained his desire for a quick set of indicators. However, during the period of the study, the national interest in SWB or happiness became much more prominent and offered a way for BVBC to do cutting-edge work that the executive officer had been keen to encourage:

> I would be really pleased if we could do something groundbreaking; there would be ten factors of happiness…the softer, qualitative, personal, individual or social feelings in an area. I don't know how better to say it than that…if I had to chose the ten things that are most important to my quality of life, or if I had to talk about my state of mind at a point in time, my happiness as I would call it, my contentment, my satisfaction with my lot.

This officer presented himself as 'the theorist in the organisation' and was generally seen as an innovator both within the authority and by members of the LSP. I felt that he was genuinely interested in SWB but it also provided him with a strategic opportunity to do something different to the county in the light of the forthcoming restructure. In any case, at this juncture, we were told by the executive director that the LSP Board was not the correct arena in which to develop this work. His reasons were that the LSP Board had too much on its agenda; it only met quarterly; and it had a limited time to discuss issues. This was certainly true and as discussed in the last chapter, there were certainly capacity problems with the LSP. However, the sustainability officer and I felt that the quality of life indicators should be central to the work of the LSP Board and, in the absence of any other democratic forum open to us, strongly pressed for their continued engagement. This struggle was to continue for the rest of the project.

The Sustainable Communities Conference

During this period, I attended the national Sustainable Communities Summit in Manchester. Speakers included the prime minister, Tony Blair, and a number of cabinet ministers including John Prescott and Gordon Brown. As an indication of how they were interpreting and promoting the concept of 'sustainable communities' the event was an eye opener for me. The dominant themes were ones of 'safer, cleaner, greener' place-making, urban regeneration with a focus on landmark architecture, 'stable' or 'sustained' economic growth, entrepreneurialism and community participation. The keynote speeches almost entirely ignored issues of global justice. In the exhibition arena there was also a focus on physical development; several local authorities were showcasing new developments or master-plans; private sector developers and planning consultants were heavily represented; and the environmental sector was primarily represented by renewable and energy-saving initiatives and eco-build construction firms. It was reassuring to see a focus on disadvantaged communities, and physical development obviously needed. Nevertheless, I was uncomfortable about the degree to which these narratives narrowed the concept of sustainability. In the evening, John Prescott, the deputy prime minister, hosted a prize-giving for the most sustainable communities. It had the razzmatazz of a game show, with jazz music, applause, and an attractive female assistant to pass him the prizes, which went to representatives of disadvantaged communities that had managed to turn themselves round. It also missed a golden opportunity to make a connection between their achievements at the local level and the possibility of making a difference on a global scale.

A highlight of the conference for me was the inspirational speech by Amartya Sen who was one of the few to put the concept of sustainable communities into a global context. He asked 'how large is a community?' and urged us to consider the 'importance of placing local wellbeing in a global framework'. Similarly, Gary Lawrence, co-ordinator of the Sustainable Seattle project, a pioneering community sustainability indicator project, observed that the idea of 'quality of life' allowed people to think in an individual way and to ignore 'transboundary responsibility'. It was at this conference that I became aware of how political agendas were framing ideas of sustainability and community. Afterwards, I became more critical of the sustainable communities discourse. This increased sensitivity also heightened my dissatisfaction with how things were going at Blyth Valley. I started to become more interested in the subjective wellbeing field and tried to understand how this could be integrated into local measurements and policy.

Differing understandings of the project

Within the first year it became apparent that there was fundamental confusion within our small team about the project and my role. The sustainability officer wanted to work with me and use the project to drive some sustainability awareness and strategic responsibility into the LSP. She saw the LSP as being the main collaborators in the work because of their role as strategic leaders in the area. The executive director wanted me to work independently from the sustainability officer and the development of QLIs, and to use the project to measure 'happiness' and conduct ground-breaking work that would be aligned with new national discourses. He saw this as being outside the policymaking process, an interesting, important, but separate study, the 'results' of which could be of use at some point in the future. He seemed to see collaboration as involving engagement with the 'community', outside the policymaking arena. The views of the sustainability officer and the executive director often differed. The strategic manager continued to focus on the need to gather statistical 'baseline data' and to maintain an open mind.

I myself had initially understood the project brief as being to develop QLIs according to *The People's Plan* and I believed it was important to raise awareness in the council of sustainable development issues. My views on participation focused on the importance of making sure marginalised views were represented at the strategic level. I did not believe that I could represent 'community' views on wellbeing, instead I saw my role as generating a limited amount of in-depth qualitative data, which might open up questions about wellbeing and indicators to inform political debate. Therefore, my preferred approach was to develop indicators through discussion and debate with the LSP, informed and supported by limited qualitative research in the local area. I was confused by the conflicting views around the project and torn between loyalty to the sustainability officer and wanting to address the interests of the executive officer. My own opinions were shifting rapidly in the face of new experiences and information. I struggled to understand my role, both conceptually in terms of the indicator project and hierarchically in terms of my position as an insider/outsider in the authority. After starting from a position of clarity, I began to feel a profound confusion.

In our meetings, there was often unresolved tension and a sense that we were talking past each other. Table 7.1 gives a typology of the different ways in which indicators were presented in these discussions.

The matrix shows ambiguity about whether the QLIs should: be focused on product or process; be objective, qualitative or subjective accounts (the latter

Table 7.1 Typology of indicator constructions

Indicator	Role of indicator	Role of LSP	Role of researcher
'Baseline data' linked to a 'State of Blyth Valley Report'.	Provide a steer for policy. Provide 'hard performance measures' to judge success.	Sign off indicators.	Information officer; collate existing sets of local statistics; identify gaps in the data.
Quality of life indicators; a limited number of 'high level aspirational indicators'.	Reflect Sustainable Community Strategy goals; steer LSP towards 'sustainable communities' agenda.	Debate the meaning of quality of life and sustainable development; collaborate in the creation of indicators.	Educator and facilitator; facilitating debate around sustainability and assisting in the collaborative development of indicators.
Quality of life indicators; representing 'community values'.	Provide opportunity for the public to participate and inform indicators.	Sign off indicators.	Participatory researcher; working with communities to develop indicators.
SWB indicators based on an academic review of evidence.	Provide objective evidence to inform policy on promoting subjective wellbeing.	Sign off indicators.	Academic reviewer; collate evidence and produce indicators.
SWB indicators based on qualitative research with communities.	Provide a 'temperature check' on happiness and contentment in the borough.	Sign off indicators.	Person with special skills; delve into the hidden lives of people and produce indicators that reach the parts other indicators miss.

two were often conflated); be embedded in or separate from an SD paradigm; be educational and awareness-raising or just provide data. How far should indicators attempt to distil a body of objective research or represent a democratically produced concept of wellbeing? Who should be engaged and how? It is inevitable that there should be different ways of seeing QLIs and their relationship to SD. As constructs they reflect deep-seated discourses and belief systems that interact in a complicated way with narratives about participation, evidence and policymaking, wellbeing and sustainability. There was a limited resource (me) and several conflicting ideas about how best to employ that resource. In addition there was increasing tension between the sustainability officer and the executive director and she became further and further isolated within the authority. After the first year, there was a sense in which the project needed a strong steer and but no-one quite knew how to do this, or whose responsibility this was. The sustainability officer was moved to another unit and withdrew from the process and the quality of life project defaulted to a series of discussions with myself and the strategic planning manager. After the first year LSP support funding ran out and the LSP co-ordinator and the information officer were made redundant. This significantly weakened the administrative support that the secretariat was able to offer the LSP.

Taking the work forward

In March 2005, central government required all community strategies to be rebranded as 'sustainable community strategies'. This offered an opportunity to bring existing strategies more in line with sustainable development thinking. As outlined in Chapter 1, the government set out eight criteria for a 'sustainable community' which local strategies were to reflect. I was asked to write a report for the LSP Board on how BVBC QLIs should relate to those produced by the Audit Commission (AC) QLIs, or whether the council should simply follow the county in adopting the latter. My report strongly advised against adopting the 'off the shelf' set for three main reasons. First, that the AC itself had recognised that its set did not adequately reflect wider sustainability goals such as global justice and responsible consumption. Second, the set was generalised in order to allow comparison across the country and was therefore unable to give an accurate local picture. For example, one of the indicators was percentage of homes without central heating. This was a proxy for reducing winter deaths. However, the Northeast has both the highest proportion of homes with central heating *and* the highest rate of winter deaths of older people. I also argued that the set excluded some important measures simply because no country-wide measurement currently existed. My final reason was that a strong disadvantage in adopting the AC QLIs would be the risk that policymakers would 'sidestep these important discussions for a convenient off-the-shelf solution which ticks all the right boxes.'[2] I recommended that Blyth Valley continued to develop their own set of indicators collaboratively. This recommendation was accepted by the LSP Board.

In the summer of 2005 the strategic planning manager and I devised a plan to move the project forward. The strategic planning manager was keen that I conduct interviews with the wider partnership, making sure I interviewed people who represented social, economic and environmental policy areas, and then collate key issues. This seemed sensible and I agreed initially. I was struck later by how powerful the SD graphic is in defining the methods for the indicator development. The assumption was that by interviewing an equal number of professionals in each of the three categories the research would be 'balanced' and would produce a set of indicators which everyone would agree to, and which would represent sustainable development. A seductively comfortable idea yet deeply flawed. Inherent in the idea of participatory democracy is debate and contestation.

In addition, people cannot be so conveniently categorised. I interviewed 31 professionals and policymakers associated with the LSP partnership and this produced an interesting and rich set of data. Most interviewees expressed complex interrelated values reflecting both personal and professional views. I was willing to collect data but argued that indicators could not magically appear out of this complexity; there had to be a period of discussion and debate about priorities and trade-offs from the issues that were raised. Otherwise, I believed, the result would be another long set of indicators to which no-one objected and which would be ignored. They would include a wide range of views but would provide inadequate direction for policymakers. One of the main tensions throughout the project came from the fact that although BVBC genuinely desired participation, it saw this essentially as a process of extraction like coal mining to supply the raw material from which I would then form indicators. I found the lack of a forum for democratic debate continually debilitating.

At one meeting, where I tried to articulate this, an officer asked why I could not just interview people and then turn the most popular issues into indicators. At the time I felt that this showed a lack of understanding of the nature and complexity both of qualitative research and the nature of wellbeing. I then realised that until one has tried such an exercise this suggestion might seem sensible. This is why I believe it is important that policymakers are more involved with the indicator process. Paradoxically, they didn't see the point as I was there to do it for them. When I tried to talk about these issues I felt as though I was making excuses for not getting on with it or coming up with the goods. I realised that although the project had begun as a team endeavour, for various reasons it had defaulted to me. This made me increasingly despondent. I began to assume the demeanour of many of my local authority colleagues, becoming demotivated, cynical and internalising problems as personal failure.

Community 'participation'

I made plans with the sustainability officer to consult with the local population. There was pressure to do something creative, visible and raise the profile of the quality of life work with residents of the area. So I decided that going out and talking to people on the streets and in venues around the borough was the best way to reach the general public, and I drew on my experience of conducting participatory appraisal (PA). With a local filmmaker I went out and asked some residents to give their views on quality of life. This was edited into a short film and (with participants' permission) shown in council offices around the borough and on the council's website. It advertised a 'quality of life week' during which the sustainability officer and I conducted public consultations at seven venues around Blyth Valley.[3] The consultation resulted in over 100 impromptu interviews with members of the public. Five main questions were used as prompts:

- What's important for your wellbeing?
- What couldn't you do without?
- What would improve the quality of your life?
- What makes you happy?
- What helps you cope with life?

The questions were designed to generate open-ended discussions on quality of life and wellbeing, and to approach the issue from different angles, using varied but related concepts and language. The questions and style of questioning were designed to be informal, relaxed and conversational so that the respondent would feel comfortable. The questions were also designed so that if people wanted they could give very meaningful and rich answers, but if they were not willing to do so, they could give a tongue-in-cheek answer. Many people were willing to spend quite a long time talking for the interview, and the depth and candid nature of many of the responses was surprising, considering the public location. Other people were happy to answer the questions routinely but did not really expand. Notes were taken by myself or the sustainability officer during the interview or immediately afterwards and these notes were typed up for analysis. The consultation was not designed to be representative nor was it intended to provide quantitative data for comparative analysis, therefore it was not appropriate to ask intrusive questions about factors such as age or social background. However, from the personal information people volunteered in talking about wellbeing, it is fair to say that a wide cross-section of the community responded.

One of the very clear aims was to avoid prior framing of wellbeing or restrict questions to domains like relationships, work or local environment. The questions were open, allowing people to frame wellbeing in whatever way they chose, including questions about future aspirations and past experience. Attempts were made to include the views of some people who may have particular wellbeing needs. I conducted eight focus groups with: the elderly, faith communities, the economically disadvantaged, ethnic minorities, young people and children.

A sustainable community strategy

In BVBC, the strategic planning manager had the responsibility of co-ordinating this new strategy and in early 2006 went to the second national Sustainable Communities Conference, again organised by ODPM. His experience was different from mine. On his return he seemed invigorated by the sustainable communities agenda. He took an 'executive decision' to use the ODPM's eight themes as the basis for the new sustainable community strategy, instead of the five themes that had been agreed locally for *The People's Plan*. In LSP meetings people seem perplexed by this as they felt that priorities for the borough had already been identified collaboratively. They also felt that this new set of themes lacked clarity; 'they seem to cover everything' was one person's comment.

Undeterred, the strategic planning manager decided to create separate teams to develop each of the eight themes. He chose eight relatively senior officers from within the council to lead these teams, and these officers formed the steering group for the new strategy. I asked him why he had not included any LSP partners in this initiative. 'Because I want this done' was his reply. However, the people he chose did not necessarily 'champion' their particular themes. In addition, the development of a new sustainable development strategy was perceived negatively by LSP partners in conversations I had after various meetings because the themes were presented to them as a fait accompli. This negatively affected their enthusiasm for the process, thereby producing a vicious circle. I also understood, though, that the strategic planning manager was over-worked and under pressure to produce a new sustainable community strategy document and applying a top-down, ready-made framework was appealing to him. He was also becoming more interested in sustainability issues and perhaps saw this as a way to bring the council more in line with SD thinking.

In May 2006, I had a long discussion with the strategic planning manager about the role of the indicator project. I was busy gathering and collating data from policymaker interviews and public consultation but I felt that unless the project was linked to strategic policy in some way it would continue to be peripheral and uninformed by local priorities. We agreed that the project should help to form the headline indicators for the new sustainable community strategy. It was agreed that I would sit on the steering group and I felt that this group would provide some sort of forum to discuss priorities and indicators, even if it remained confined to council officers. Hopefully, during the process, a greater spread of consultation would occur. However, weeks passed and no steering group materialised. It became clear at this point that things were being put on hold due to the Local Government Review.

Local Area Agreements: another new initiative from central government

As explained in Chapter 5, in 2006 the government introduced another layer of local governance reform, Local Area Agreements (LAAs). NCC was one of the councils selected to pilot this initiative, and there was an intense period of activity to realign priorities towards the LAAs. The government made it clear that the LAAs were the delivery mechanism for the community strategies. The NSP and all the district LSPs had to sign up to action to deliver mid-term outcomes in 3–5 years, and if they did not, they would have greater flexibilities taken away. At a local conference to inform officers about the LAA process, I sought clarity from policymakers and researchers from NCC about how the QLIs would fit into the

LAA structure, but no-one knew. One senior policymaker felt that 'people are too switched off on process, they need action and the LAA will deliver this'. However, other policy actors had mixed views and felt they were steered to choose indicators that they knew they could improve on or which had finding attached. This limited the scope for local indicators to be effective and narrowed policy gaze:

> Local indicators can also be added to the total of 53 chosen for an LAA, but it will be hard to put funding against them. Given what we know about statutory performance indicators, if you don't have to measure them, and you don't get any extra money for other locally identified priorities, then you are less inclined to do anything.
>
> (Policymaker)

> The Local Area Agreement seems to have almost narrowed things down rather than created a comprehensive approach...There's no targets in there around teenage smoking, alcohol consumption or obesity.
>
> (Policymaker)

The sustainability officer and I were concerned about where QLIs would fit into LAA indicators, and about the capacity of a struggling LSP to cope with yet another new initiative and set of targets. The LAA contract drew on the energies of many officers working at strategy level, and there was little interest in discussing how quality of life indicators would fit into this new structure and the sets of indicators produced. QLIs seemed to drop off the agenda and this felt to me like the last nail in the coffin.

Product pressure

In April 2007, I was due to present a final report to the LSP Board on the indicator project. Initially the sustainability officer and I worked on this together. We felt pressure to come up with a set of indicators but found this very difficult to do considering the lack of clarity. We lacked any firm steer from the council or the LSP about what or who the indicators should be for, or how they were to fit into policy. Northumberland County Council were at the time consulting with all the districts on a draft set of Local Area Agreement indicators and also BVBC was supposed to be producing a refreshed sustainable community strategy. The sustainability officer tried to get advice about how QLIs fitted in to either of these processes but no-one seemed clear. We were torn regarding whether to try to influence either (or both) of these processes by trying to fit around existing indicator sets or by coming up with a new set completely. We also wondered about not coming up with one at all as we didn't see the point.

We looked at the data from the public consultation and focus groups and compared it with various sets of indicators. After two days we were finding the process frustrating. The sustainability officer gathered the LAA indicator sets and other sets of indicators and we spent two days trawling unproductively back and forth between these sets and the data. We found that different sets of indicators were underpinned by different agendas and so it was difficult to make direct comparisons. For example the BVBC data largely focused on a broad conception of (mainly) individual wellbeing and the other indicator sets were focused on an area-based concept of quality of life or on sustainable communities. Our data were pointing in a different direction and so existing indicators could not just be plucked from other sets and applied. In addition we still hadn't agreed or identified an underpinning concept of quality of life or what the indicators represented.

We had long since lost the initial clarity we had about the purpose of the project and I personally felt completely swamped with sets of statistics and indicators none of which 'spoke' to each other. At some point, I just got angry. It wasn't my role to choose indicators! These decisions are political! This was supposed to be a collaborative project! How can one person do this! How can I legitimately make decisions about value conflicts and tensions in the data? Therefore I did what was in my capability to do and I collated and presented the themes from the qualitative data and tried to provide a critical account of the problems and gaps around current indicators and what they needed to be effective, including tensions surrounding wellbeing and sustainability understandings. I strongly resisted the pressure to produce 'gloss', in this case, a neat list of indicators which looked impressive but didn't actually reflect any democratic process. I argued that the choice of indicators, the choice of how we measure wellbeing, was a political project involving debating priorities and trade-offs. I was well aware that some would see this as an excuse for personal failure and I found it hard, despite my deep convictions, not to personalise this. Discourses of 'hard evidence' and 'rational driven policymaking' were powerful and affected the way I valued my role.

The culmination of these efforts was presented to the LSP in April 2007 as a draft report.[4] It outlined seven areas of wellbeing and illustrated these with qualitative data from the interviews, focus groups and public consultation. The findings of that report form the basis of the next chapter.

Conclusion

The process of trying to develop quality of life indicators went through a long and difficult journey, hampered by many institutional obstacles. There was a broad struggle between using indicator development as a *product*, developed outside the policymaking arena, to directly influence policy reflecting an evidence-based rational view of policy; and using it as a vehicle to enhance partnership working, understand conflicts, making values, norms, priorities explicit, where the focus was not only on developing content (what is quality of life and the measurements for that), but also on reflecting on the *process* of governance. However, this was not so clear cut as officers who promoted the first approach were very keen for community involvement but 'out there'. The sustainability officer, however, was less interested in community involvement (although she certainly did not devalue it) as her main focus was on working with the LSP and senior officers and stimulating debate at strategy level. In adopting the second approach, the sustainability officer and I became process heavy, and made demands for time, debate and engagement at the highest levels. In hindsight this was unrealistic as we did not fully recognise that a lack of resources, and the district/county split were undermining the capacity of the LSP to deal properly with this project.

Those who inclined to the first approach, put pressure on the work to be somehow scientifically robust and seemed to think that this was a case of interviewing people and doing some sort of qualitative/quantitative exercise to determine which were the most popular answers and then use these to come up with a definitive list of indicators. The rationale was that somehow through an 'objective' process of gathering equal amounts of views on the three domains of SD (or various domains of wellbeing) then you would automatically arrive at the perfect SD (or wellbeing) indicator set. This requirement to be 'equal' also reflected the desire to 'solve' tensions between the environmentalist and regeneration officers, the implicit assumption being that such tensions could be dealt with by a 'fair' splitting of the pie whilst avoiding the need for conflict. The requirement on one or two officers to do the deciding on the indicators also avoided messy discussions and 'talking shops'. I believed it was unfeasible to produce valid indicators through a mechanical approach to this data, and that this would close down democratic debate rather than engage with different viewpoints.

The indicator project took place in a continually shifting landscape that was heavily influenced by personal agendas and institutional discourses, a lack of strategic partnership and, crucially, the lack of an effective democratic forum in which to debate wellbeing definition and measurement. It was clear that the sustainability officer and the strategic planning manager had responsibility for a heavy and complex workload without sufficient capacity or structure in either the LSP or the council to support them. These factors meant that it was difficult to develop real clarity about what the indicators should reflect, who should be involved in creating them, or how they should be integrated into policy and practice.

8   Developing a wellbeing framework

You can't measure something when you don't know what the definition is.
(Pupil, age 14, Blyth Valley Community College)

This chapter describes the work of producing an effective concept of wellbeing to inform strategic policy in Blyth Valley. In the first instance, the focus was not on measurement, but on constructing a local account of wellbeing, in a participatory, non-prescriptive way. Although the aspirations for a collaborative indicator project faltered due to the issues described in the previous chapter, I nevertheless conducted a series of seven public consultations (involving over 100 people), eight focus groups representing different interests, and in-depth interviews with 31 policy actors from the local authority and the wider LSP partnership.

A 'wellbeing framework' with seven main themes was produced from this qualitative data (see Table 8.1). I discuss the nature of this framework and the choices and tensions in producing it. I then compare this with other accounts of wellbeing, particularly Martha Nussbaum's list of 'Central Human Capabilities' and the Audit Commission's set of local quality of life indicators. I carried out a mapping exercise which revealed some interesting commonalities and differences between these two accounts and the Blyth Valley account (see Appendix C) and this provides the basis for a discussion on the political choices involved in producing local wellbeing indicators. The challenges of transforming a complex body of qualitative data into a coherent and widely resonant account of wellbeing are legion. What is presented below was a first step: I therefore include some pointers for further work.

Understandings of terms and concepts

Chiefly I wanted to avoid either adopting a pre-emptive framing of wellbeing or limiting my questions to standard domains such as relationships, work and local environment. I used open questions to allow people to frame wellbeing however they chose, and included ones about future aspirations and past experience. I used the terms 'wellbeing', 'quality of life' and 'happiness' interchangeably to try to capture a full range of meanings. If possible, I asked people to clarify what the terms meant to them. A few respondents thought quality of life and wellbeing were the same thing; more commonly they believed them to be different concepts, although some struggled to say why. Many people associated quality of life with environmental or socio-economic factors, and saw these as providing the right conditions for wellbeing, which they described as a state of individual health, happiness and fulfilment.

> Quality of life is how I'm living, wellbeing is how I feel about myself.
>
> (Public consultation)

> Quality of life is what you do and the choices you make, wellbeing is inside and outside health, inner self-esteem, motivation, healthy diet.
>
> (Public consultation)

> Wellbeing is, like, you. Quality of life is, like, what you're given.
>
> (Pupil, age 15, Blyth Valley Community College)

> And quality, quality is quite a subjective term, but quality of life has a direct impact on…people's wellbeing.
>
> (Policymaker)

> Quality of life could be the processes to create the output which is wellbeing. Like, wellbeing is happiness, healthiness, those kind of things, and the varying degrees of those are determined by the quality of your life, like where you live, your job…the environment, your friends.
>
> (Policymaker)

On the whole policymakers and residents recognised that the local authority had a role in improving quality of life. The picture was less clear for wellbeing, which was also linked to individual personality, circumstances, family background or the natural ups and downs of life. In general conversation people often used 'happiness' and 'wellbeing' interchangeably but when discussing the detailed dimensions of wellbeing they usually viewed happiness as a component of it.

The seven themes of wellbeing

Much of the discussion about wellbeing in Blyth Valley highlighted its multi-faceted quality and the interdependent nature of its components. It was necessary to tease out these different strands to make a coherent picture that could be communicated easily to the public and policymakers. I believed it would be misleading to reduce the complexity of wellbeing into populist sound bites (which can sound patronising and prescriptive) but I also wished to ensure the data would be readily accessible. I was clear that accessibility did not mean consensus: while it is possible to view the framework as a coherent account of local wellbeing, there existed tensions both between certain aspects of wellbeing and between theories about how it could be achieved. I discuss some of these further below.

The public consultation responses were analysed first, by taking account of everything that was mentioned, being sensitive to the context in which issues were discussed and the links with other aspects of wellbeing. I did not prioritise issues according to the frequency of response as this might disadvantage minority interests. However, how often an issue was discussed did influence the categorisation of wellbeing themes; I made large clusters of similar issues into themes and organised smaller clusters as sub-themes round them. Categories in themselves can suggest priorities, and are open to various interpretations. There were a number of ways the data could have been organised, for example some people cited faith as being important to their wellbeing. One option would have been to create a separate category of 'spiritual wellbeing'. However, I was not sure how prescriptive this might seem, as many people did not mention spirituality. In addition, many of the people who talked about their faith also mentioned being intimidated or victimised because of it, and so I chose to include it under a category dealing with 'freedom' – the ability to live the life you choose without discrimination. I was keen to have discussions about these sorts of issues in the further development of the work because I felt there were many decisions that needed to be taken collectively.

I organised the data into seven themes: personal qualities; health; activity; income; social relationships; environment; and freedom. Members of the public spoke to me on a one-to-one basis, so the data from public consultation reflects a series of individual views. But it reflects more than ideas about individual wellbeing; subjects such as social values, fairness, and reducing social disadvantage were discussed in depth, as were the institutional arrangements needed to bring them about. The focus groups added detail and nuance to particular issues. In contrast, policymakers were inclined to talk in terms of strategy unless specifically asked about their personal wellbeing. Many, for example, tended to discuss raising local employment levels, whereas members of the public might talk about job satisfaction and the value of leisure activities. Although they viewed 'employment', 'education' and 'housing' as being important, they did not present these as primary themes. Instead, their narratives of wellbeing reflected their view of the world and their sense of themselves in terms of who they were, how they spent their time, how they related to people and places around them, and how they managed or struggled to cope. I therefore chose to organise the wellbeing themes accordingly (see below). I particularly separated 'income' from 'employment' whereas these two categories are often conflated in indicator sets. The data from Blyth Valley show clearly that both the quality of income *and* the quality of job/activity is important and more attention should be paid both to these aspects of wellbeing and to the distinction between them. Although the analysis included the views of both policymakers and the public, its categories and language predominantly reflected the discourses of the public consultation and focus groups.

It is of course impossible to make a neat separation between the public and policymakers, many of whom not only lived there, but had also grown up in the area. While some policymakers drew a sharp distinction between their own lives and the very different needs of disadvantaged people in Blyth Valley, others spoke about their personal quality of life and how that influenced their work. So rather than concentrate on public responses as if they represented a pure view of 'local' needs, I made the integration of policymaker views explicit, and highlighted where they diverged significantly from those of the public.

In addition, the ordering of the data into these themes was not driven by philosophical categorisations such as needs, capabilities, resources et cetera. Some of the factors which people felt were important to wellbeing related to subjective wellbeing, others to needs, capabilities or resources; while others related to institutional arrangements and structures. I resisted the urge to fit the material into these categories without further deliberative and conceptual development. I chose to present an account of wellbeing as it was expressed and understood by those I spoke to, rather than to seek philosophical purity.

Instead of presenting a mechanical account of the whole range of issues in each theme, I have selected five voices (below) to illustrate the complexity, interconnectedness and tensions in creating a single account of wellbeing from qualitative accounts of individual lives. The data is summarised in Table 8.1.

Person A

I had one teacher, he was a maths teacher, and I can still see that man. I thought I couldn't count, he had me like this [shows trembling hand] and if you got one sum wrong you got a whack for it, if you got ten wrong you got ten whacks. And I can still see that man writing on the board…it wasn't until I left his class that I learnt that I was quite good at maths. So I don't agree with physical punishment to that extent. My mother never touched us, never once. This is why I say it comes from the cradle.

There's not the material poverty now, not to the extent that we knew it. I remember when I got my first fitted carpet and a telephone and I'm sitting thinking I've made it, I've done it. But now they start with all of that, they have all of that before they get married now. We got one thing at a time and saved for the next one. When you think about it, materialistic wise, I bought my first house and I paid sixteen pound and six and ha'pence a month mortgage. My mother was mortified that I was buying a house. However, now, it's two hundred thousand…

Well I went out to work for three months, it lasted twenty-three years. I was told in a lecture that children of working parents were deprived. So I said to [name of child], I have to ask you a question, I've been told at college today that children of working parents are deprived. 'Now', I said, 'how do you feel?' He says 'Mam just you keep slipping me the odd pound and I'll not mind' [laughter] and he was never deprived by that, but I had a surrogate substitute, you know, my mother was there.

I feel very sorry for young parents. I do think they have a much harder time than we did because we respected our parents and we were controlled to a certain degree and we didn't do things we knew we shouldn't do because we didn't want to upset our parents. I do feel it's a much harder task for the

Continued …

parents now to say no to the children, because of the peer group pressure, you know, the designer this, the designer that, and it's hard for the parents to have to cope with it. And the children are living in this society of materialism and you know what it's like if you are the odd one out. It's still got to come from the family but you know it's not easy to always say no.

Shall I tell you about when I retired, knowing that when you retire you haven't got to just vegetate? I phoned from my office and I said to the college do you take geriatric students? And he started to laugh and I said I'm serious. Now that was eleven years ago and I'm still going. I've met a lot of good people, done a lot of things in eleven years and it was all through the education that was offered. They saw me as a person not as an old age pensioner. I worked with the elderly and nothing infuriated me more than someone would come in and say 'they are all sitting in the same chair and they all do this', I says 'listen they are people, you know'. People categorise, pigeonhole people…

The thing for this age we are at is our attitude, there's lots available if we are able to get out to do it. The groups that might find it more difficult are the ones who are immobile, those that need additional support…This thing we are talking about, our health and our wellbeing and people who can't get out into the community, there's one point that I really would like to raise is the fact that they are closing the libraries left, right and centre and that is wrong. And I do feel very strongly about that, that it should be more available than less. They should have in the library, I've always said, a coffee room so that people could go and meet and converse, have time to pick their books, discuss what books they like and advise other people you know what's available, what's not available.

Wellbeing is health, physical needs, emotional needs, psychological needs. Quality of life, which is linked with wellbeing without question, I think your quality of life can still be good if you are frail and elderly but I do feel that your quality of life is controlled by your ability, your attitude, your whole perspective, your life experience and everything, freedom to have the mobility to get around and make your choices.

Person B

In those days [name of place] was probably considered a very close-knit community: everybody helped everyone else and knew everybody else. There were no secrets in the village. A great deal of community feeling about the place and I came across one of these books that have been written about [name of place] and I was just amazed at the number of societies that there were…so they created their own entertainment. The shops survived and there's a great variety of shops because people would buy things locally, shop locally. They would go to the pub, and there was a couple of pubs, a social club. And their entertainment was in the local institute where, yes, some really good memories of a cinema, the film coming on. But our entertainment in those days was going down to the local dene with the river flowing by. And during the summer holidays, yes, there's a true story, you could write a book about it, there was about half a dozen or so in the gang, and we would go off from light and come back at dusk. And your parents with no mobile phones in those days…and there was a great community spirit and you had the local policeman…In those days our holidays were few and far between and we never went abroad, that's for sure. But parents in those days, bless them, everything that they made they gave to us to pay for this, that, and the other. So they sacrificed a lot for us.

My quality of life is, I keep saying it to her, she doesn't believe me, having a really great wife, great soul mate, great partner and having two fantastic kids and I couldn't have wished any more. When the [children] were born she stayed at home, looked after them, took them to school, brought them back from school, which was her choice, I wouldn't have pushed her one way or the other. And when you get older, your quality of life very much perhaps relates to your partner, your family. You think more about, 'Well, what future are they going to have?' And I just think, about my two children…you just think maybe they don't realise what's around the corner in terms of climate change.

Happiness, family, environment; I suppose things like peace, just a peaceful world. Tolerance: I used to be a real tearaway when I was younger, but I still had a lot of respect, I still had a lot of respect for older people, believe it or not, maybe because we had a village policeman and if the village policeman told my father, you know, I was in the doghouse. There is a lack of respect from both parties [now], from the older ones about the younger ones and vice versa. So there needs to be, I think, a little bit more integration.

I enjoy travelling now, now that the children are just about gone – I suppose the next stage is helping them to purchase houses and what have you. We're not at a bad stage in our life in terms of having a bit of disposable income such that we are able to go on EasyJet flights and so on and that's been great, because I never had the chance when I was young. Mind you, I think quality of life, I always think life was better in the olden days. People were more caring in general terms.

Person C

For me it's a very personal thing, I think it's about being free to make the choices of how you live your life, and within the constraints. I think it's unrealistic to think there's going to be a perfect way of life, so everybody has constraints, but it's how many choices you have to get out of some of the problems. So there isn't an ideal life but there is a way of choosing the best choices for you, the best path for you through life.

Continued ...

Person C continued

Certainly I think my choices were informed by education, education was a huge factor in my life. I came as an immigrant, when we moved here it wasn't multi-cultural. We were practically the only non-white faces in the area and the way that we got out of poverty, racism, discrimination was through education and we didn't have any choice but to accomplish at school, get a good education and that would lead to all sorts of doors opening job wise. So that's what I did, I went through a quite academic route. It's education every time, not the education system but the fact that from birth I don't remember a time where education wasn't important to me as something that was, that I could change my life through. The way I was brought up and the values I was given as a child and the evidence of my eyes. And that's still true now, if I stop learning then I've actually stopped having any quality of life.

And the other thing has been health, I've been a healthy person all my life. So those two things together have given me some life chances that have really made, you know, I've got a good life. We all go through our ups and downs but those two things really, if I hadn't had them, my life would have been totally different and for me that's, I mean you know, it's not even money, it's not even about money, although I do think poverty is one of the scourges of life, you know, you can't do anything. But it's not money, you can have money and still be quite poor in all sorts of ways in your life but your education and health can overcome that poverty. So I think for me that what's made a difference.

I'm a confident person, I've overcome a lot of prejudice previously but you still hit it and it takes away all your choices. It wouldn't matter if I was a brain surgeon on a million pounds a year I would still get the name calling and that has affected my quality of life because that's not something anybody should deserve because of how they look. But also it influences how you perceive your fellow citizen because you know, you think, they're probably quite an OK person but they feel it's OK to shout abuse at someone who's not like them. So that makes me angry towards that person and I shouldn't have to feel that anger towards that person. So prejudice is probably the biggest thing that's hindered me.

What I remember about close-knit communities was they felt free to shout abuse because I didn't belong and I was a different colour. So I'm very wary, I don't have that nostalgic view because my background was very different. Having said that, when thinking about children, schools were the leaders, weren't they, in their communities, they were the ones who were respected in the community, if a teacher said your child is misbehaving. Now they are being challenged all the time by parents. But there were other leaders in the community. The church is dwindling, although I'm not a religious person and I certainly wouldn't push dogma, but they were beacons to say, this is a way of behaving. I suppose what we've got to do is identify the new leaders. What used to happen, you know, from my perception was that the church and the school and the police to a certain extent came from that community, they lived and worked in that community, so that's not going to be possible any more because we've moved on for all sorts of reasons. Perhaps there are institutions that have to take a leadership role…I don't know, I think it's a conundrum.

We made a conscious decision to move to live near the sea. I was brought up on a coastline so just the drive to work always lifts my spirits, I can face a hundred different problems coming to work and a hundred different problems going away from work but just being able to go to the beach and see the sea has just been fantastic for me.

Person D

It's sort of how you measure how good your life is, how comfortable you are with your surroundings, how satisfied you are. I like being outside, it's more interesting than inside. It's always changing. I like walking and being with people I enjoy being with, sitting and talking, debating, arguing. I don't like being on my own. School is boring though.

A lot of people don't like telling people they are Christian. It's got me a few beatings, but that's part of being a Christian. I don't really mention it, even if people don't have a problem with it.

There's a lack of facilities for young people, there hasn't been a cinema since 1939 or something. It's more important to have a place to meet than an activity. But if you put on a place to meet it would get colonised by charvers[1] and then get basically trashed. I usually avoid the places where charvers meet, I feel intimidated. I had a paper round and I stopped it because the charvers were giving me grief. They were shouting 'you freak, I'm gonna smash your face in' and throwing things at me. That really annoyed me, that I had to stop. I couldn't go to the authorities because they hadn't really done anything and you can give them ASBOs and stuff but that doesn't really work.

I can't remember charvers when I was little, now it's huge. It probably started because there wasn't much to do. Now with computer games you don't have to be able to think of games to play. Kids of our age don't have to invent things to do. But it's not just that. I think it sometimes can be the way things are at home, a lot of them come from quite dysfunctional families. They are trying to find friends outside that. I spoke to a charver once and he said he was terrified at home and the only way he could survive was he took it out on the streets, all this rage and aggression.

Divorces have become more common, more single parents and broken up families. Some young people have realised that they have more freedom, they've realised that we live in this massive bureaucracy and they can probably get away to Newcastle and parents won't know. They have the ability to travel whereas years ago people stayed in their own area and people saw you.

Person E

Being happy, having love, having friends, having a decent home with hot water, some people don't have it, stable job, being respected, enough money. I think as long as I'm happy that's the main thing. I don't think happiness has got anything to do with money, well, there's a relationship but it's not the main thing. You just need enough to get you everything you need and a little bit left over. Material things shouldn't be important but they are, that's just the way it is now, it's like as much as it's a pain, that's just the way it is in our generation now. People always seem to go on about it as if you are shallow but that's the way it is now.

There's not very much to do because you are usually just wandering around. There are only places where you need money and we haven't got any. There should be more social areas off the street, somewhere warm, somewhere you don't have to pay to get in. Lots of charvers are catered for in the community centres, activities like MC-ing and new monkey music, they voice their opinions more than we do. They used to have girls' night at [name of local youth centre] where you could go and watch videos, that was really good, but everywhere is full of charvers. I wouldn't want charvers to be there, but it's not very fair to have a sign saying 'no charvers' but if you are in a place and charvers are messing about, you are scared.

It's unfair because at school, you could get a student that was working mint all year, they get nothing. You get someone who is really horrible all year round, they do one good piece of work, they get loads of awards chucked at them, I don't think that's fair at all. Charvers disrupt the class and teachers spend too much time being distracted…all young people get stereotyped because of charvers. Older people don't trust us.

If there was more discipline people wouldn't be going out at 14 and drinking. Parents let them get away with anything, their mams don't care. They don't respect the police or authority. There's no consequences. I saw a girl getting beaten up the other day at the bus station. It was really horrible and no one was doing anything because it was charvers, and I wanted to do something but if I did something then I'd get into trouble.

You need support and encouragement from your parents, a lot of it comes from your parents, unless they push you to do something different you just end up being like them. A lot of teachers just tell you your bad points; you need parents who boost your confidence, don't knock you.

I don't think Blyth is very friendly, it's intimidating. I don't think anyone talks to each other any more, maybe family and next-door neighbours but people don't go out of their way to make friends. The minority stand out because of their views: in our street there's a lesbian couple with a baby, they are not very accepted, people don't bother with them but I don't think anyone would do anything extreme or anything.

> A lot of people want to get out of here, like when you go to uni, and not come back. I would come back if more people cared for it. The council clean it up but then people just trash things again. There are some areas of Blyth that people trash, they get fixed and there's too much emphasis on those places. People just wreck the facilities all the time.
>
> I like it that Blyth is near the sea, that's pretty cool, not many people could say that. I like the northeast, there's more countryside. We went on this trip and there was a load of big forests and I loved going in them.

Inviting people to speak about wellbeing in the context of their everyday lives inevitably created discussions of a wide range of interconnected issues from personal to social to global. It also highlighted how wellbeing needs are dynamic and change over time, not only because of different stages in the life course but because life has its inevitable ups and downs. It may be possible to cope with the lack of some aspects of wellbeing over the short term but in the long term many aspects are co-dependent. For example, economic security and relationship stability were often interrelated: loving relationships and family support were perceived as essential in helping people through financially tough times but long-term debt was seen as putting a heavy strain on relationships and undermining trust between family members. While there was broad consensus for some issues, there were significant tensions around other issues. One major area which was often discussed was the role of community.

Community spirit was a strong theme and this was related to a common belief that society was becoming more selfish, individualistic and protectionist. Many people reminisced about their childhoods, when life was harder but people were closer, happier, and made their own amusements. This idea of community acted as a hold-all for many other ideas, some of which are both means and ends of wellbeing: long-term relationships of support; belonging and identity; less traffic and reliance on cars; more equality and trust; fewer material possessions. Key policymakers often valorised the strength of the 'old mining communities' in Blyth Valley for their social support and seemed to want to recreate aspects of that, while acknowledging a changing world. However, others were cynical of this 'nostalgic' view of close-knit communities: 'If you didn't belong, you were demonised.' Every focus group reported experiences of discrimination or intimidation, sometimes serious and criminal, whether to themselves or friends/family members. Policymakers who were women and/or gay and/or non-white also talked about this; one policymaker described the reaction of local people when they discovered she was a lesbian: 'Being attacked and abused, hate-stuff and actually being driven out of where I lived because of it.' In contrast, the awareness of key policymakers, who belonged in such communities as children and had little personal experience of such discrimination, seemed low.

Being able to trust a good neighbour and ask them for help when needed was rated as being important by most, whereas the idea of a strong neighbourhood

Table 8.1 Blyth Valley Wellbeing Framework

Wellbeing theme	Basic descriptors (for a fuller account see Appendix C)
Personal qualities Who we are	Positive/thankful attitude to life Philosophical approach/realistic expectations Sense of humour Inner peace/self-knowledge Being happy/content with life Emotional resilience/adaptability Self-esteem/confidence Initiative/motivation Being 'other-regarding': honest, respectful, caring, understanding, sharing, generous
Health How we are	General physical fitness and exercise Healthy diet and healthy weight Reducing serious illness Alleviating suffering caused by chronic pain/long-term illness Drug, smoking, alcohol reduction Mental health Opportunity to rest, relieve stress and recover in natural/tranquil surroundings Support for elderly Support for carers Quality of GP and health services Quality of death and support for those bereaved
Activity What we do	Keeping active and busy Ability to pursue interests/hobbies/play Provision of activities/facilities Ability to enjoy 'simple things' Availability of jobs Quality of job Quality of education
Income How we manage financially	Adequate income Economic security Control over debt Fair distribution of wealth
Social world How we relate to each other	Loving relationships with family/friends Ability to care for others/opportunities for voluntary work Community spirit/compromise General friendliness of area Collective parenting/boundary setting Good intergenerational relationships Mutual interest societies
Physical world Experience of our surroundings	Clean water/air/land Tranquillity, peace and quiet Beauty and diversity (of built and natural world) Opportunities to experience the natural world Comfortable, affordable homes for all Housing security Clean, peaceful, safe neighbourhood Access to facilities and services Waste disposal and recycling Transport and mobility Renewable energy/reduction in fuel poverty Reducing climate change
Freedom/choice/ control How free we are	Being able to live the life you choose Freedom from discrimination/intimidation Fairness in how people are treated/equality of opportunity Participation in and influence on decision-making Access to support, advice and advocacy

community was more conflicted. The quality of social relationships seemed more significant than their local structure; this came out in discussions around the 'balance between support and privacy', with some being 'concerned about gossip' and the over-bearing nature of some communities where they have more contact than they would prefer with neighbours. Being able to access networks and support outside the immediate neighbourhood was important for some, so mobility was regarded as essential, and this conflicted with the common desire for general reduction in car use, for both social and environmental reasons. Significantly, what also emerged clearly was the importance of chance meetings and spontaneous chats, especially for those living alone; being able to bump into acquaintances or passers-by in the street or town centre, and the sociability in an area: 'interactions with people generally, having a joke with shop assistants'; 'random acts of kindness'; 'the kindness of strangers'. One policymaker said she hated the word 'tolerance' and that she was brought up to show 'hospitality'. This related to a significant theme in discussions about societal values. Again this was a conundrum, as, for many, it related to social norms instilled by parents or strong community figures to create a respectful and caring society and yet goes back to similar issues rehearsed above about inappropriate or restrictive norms being imposed.

Another interesting and linked tension was highlighted by the striking difference between policymakers' discourses about social equality and the public, who tended to talk more in terms of fairness or injustice. The public often mentioned fairness which they equated with addressing specific injustices and rewarding those who deserve it. There was a marked difference in the accounts of policymakers, who gave more weight to the ethic of equality and promoting social justice more broadly. For example, children talked about the unfairness of being reprimanded for things they had not done. Young people (as person E above) complained that while they worked hard, others who were usually disruptive in class were praised comparatively more highly when they did something well (perhaps by teachers attempting to redress perceived inequalities in family background). The issue of 'charvers' loomed large among the young people (from all social classes) I spoke to. They talked about how unfair it was that young people were stereotyped, inconvenienced and intimidated because of the behaviour of a minority, who seemed to be rewarded, although they recognised the complex reasons for their behaviour. This issue was under-discussed by policymakers, who focused more on facilities for young people generally, young people's health and equality of opportunity (focusing again on the disadvantaged).

It is important that when looking at the summary of wellbeing themes in Table 8.1, these differences are borne in mind. The table represents a large body of complex data and is necessarily abridged; for more detail in each theme see Appendix C.

Comparison with Nussbaum and Audit Commission accounts

My study at Blyth Valley was driven by local people's views on wellbeing and quality of life, in order to gain an initial understanding of what was valued before we began the work of developing indicators. I compared the seven themes above with Martha Nussbaum's set of Central Human Capabilities, which is a universal social justice approach. I also compared it with the Audit Commission's local quality of life indicators, which is a measure of a neighbourhood-based concept of 'sustainable communities'. This was a challenging process because all three accounts use different constructs of wellbeing, making direct comparison difficult. In addition, two are accounts of wellbeing while the AC indicators are a measurement tool. However, this is also what made the mapping interesting as it highlighted particular, although not wholly unexpected, tensions between different underlying constructs of wellbeing, and between ideals of wellbeing and measurements of it. The mapping exercise is shown in Table 8.1 and I discuss it below.

Nussbaum's central human capabilities

Nussbaum has set out a set of 'fundamental entitlements' (Nussbaum 2003: 34). For Nussbaum (1999: 70), 'The core of rational and moral personhood is something that all human beings share, shaped though it may be in different ways by their differing social circumstances.' She believes this should be the basis for 'defining the public realm'. A similar approach was attempted at Blyth Valley, to form an initial account of wellbeing in order to inform a political conception of the good as a basis for development; this was not derived philosophically like Nussbaum's, but empirically, by talking to local people and, as emphasised above, this was not a finished project. Nussbaum's set of 'Central Human Capabilities' are:

1 Life
2 Bodily health
3 Bodily integrity
4 Senses, imagination and thought
5 Emotions
6 Practical reason
7 Affiliation
8 Other species
9 Play
10 Control over one's environment (political and material).

(For a full account of these entitlements see Appendix B.)

The capability approach holds that a purely resource-based set of measurements is inadequate because people need the capabilities to access those resources. Nussbaum therefore excludes those aspects which are purely resource-based. For example, notable absences from her account are income and some environmental conditions. Allowing for these, the Blyth Valley wellbeing framework maps relatively closely onto Nussbaum's account. They both reference emotional and mental development and reflect the many ways in which activity and self-expression are important; not just *what* we do but also the quality of experience.

A striking commonality is also the way in which both accounts emphasise the liberty to lead one's own life and freedom from intimidation. Ideals of freedom and justice are strongly represented in each.

In the Blyth Valley account, the public stressed fairness rather than social justice. For Nussbaum social justice is defined by a minimum threshold. Fairness, however, is a relative concept and depends on comparison between people; something is only fair or not fair in relation to something else. Social interconnectedness, culture, values and context are at the root of this concept, Nussbaum's however, depends on a universal threshold at which each individual achieves a social minimum of justice. This is an important distinction. How can we deliver social justice whilst still being perceived as fair? Can some perceptions of fairness work against social justice policies?

Nussbaum does not include positive affect or happiness as her focus is on capabilities, however these were commonly mentioned in the Blyth Valley material. The two accounts also differ in that Nussbaum's contains no reference to the ability to receive care when ill, infirm or in need of long-term support. It does however refer to the ability to have good health and to avoid 'non-necessary pain' and her ideas about affiliation may include an implicit assumption of reciprocity. I felt her account did not measure up well from the point of view of a carer who, while not directly disadvantaged, might find it emotionally difficult to cope with the demands. I tried to put myself in the shoes of someone struggling to cope with an aged partner with dementia, as an example, and I found that a reading of Nussbaum's list spoke strangely *past* me rather than to me or against me. There would be little in her list that I could object to, but neither would it offer a guarantee that I would receive support. Similarly, she says nothing specifically about the quality of death, although her focus on planning one's own life, dignity, non-humiliation and the avoidance of non-necessary pain may address this generally. Once again, these issues are perhaps highly sensitive to cultural norms which she says must be supplemented by local accounts.

As previously mentioned, Nussbaum has been criticised for including a capability for a relationship with other species and the natural world, but I found this capability to be highly resonant with the Blyth Valley material. Many people described how access to nature (in a variety of ways) was not only beneficial to their wellbeing, but a fundamental part of it. The ability to appreciate the beauty and peace of the natural world, to hear birdsong, to rest and recharge, to walk and gain exercise were often mentioned.

All in all, I felt there was much in Nussbaum's account which spoke to the Blyth Valley experience. However, I felt that her account, focusing as it does on the individual, lacked an explicit recognition of the interdependencies between people, although she does recognise the role of social affiliation, respect and institutions in supporting individual wellbeing. And although I understood the conceptual reasons for omitting income, environmental factors and wider social interconnectedness, I began to feel that her account was too 'stripped back' to fully resonate with the Blyth Valley empirical material, which included the importance of things like general friendliness and trust in society, fair distribution of wealth, environmental quality and security. However, I must acknowledge that Nussbaum has always maintained that her list provides the social minimum of justice and should be supplemented by local accounts; what she finds problematic is if the local account does not allow this social minimum of justice to be attained.

Audit Commission's local quality of life indicators

As discussed earlier, The Audit Commission (AC) produced a set of local QLIs for use by local authorities. The set focuses on 'communities' or 'neighbourhood', in contrast to Nussbaum's focus on individual wellbeing. The AC set was driven by the government's narrative around 'sustainable communities' and is underpinned by the development of comparative measurement between local authorities. It is largely orientated around indicators for which there exists well-established national measurement. It consists of 45 indicators, themed as follows:

1 People and place
2 Community cohesion and involvement
3 Community safety
4 Culture and leisure
5 Economic wellbeing
6 Education and life-long learning
7 Environment
8 Health and social wellbeing
9 Housing
10 Transport and access.

A subset of 'aspirational' indicators was also produced for which at the time of the study there was no national data set. (For a full list of these indicators see Appendix A.)

The AC set is limited to the 'local area', and is predominantly a provision-based approach reflecting the perceived remit of local authorities. The AC set seemed to be inadequate in comparison with the Blyth Valley wellbeing framework in terms of a number of omissions. The predominantly local focus seemed to reflect the residential experience rather than residents' sense of their life as a whole, including the private, work and social spheres. Although the AC set touched on some aspects of community relations, it offered little of real substance in terms of reflecting the crucial importance of social relationships and societal values.

Although the 'local' focus of the AC set is understandable considering the concept of sustainable communities underpinning it and the fact that it is related to local authority provision, even in its focus on community cohesion it does not adequately reflect the Blyth Valley data. The AC set regards community cohesion primarily in terms of race and religion, rather than gender, sexuality, disability, age, body shape etc. However, respondents in Blyth Valley recounted experiences of discrimination or intimidation on all these grounds. In addition, the indicator measures 'the percentage of residents who think that people being attacked because of their skin colour, ethnic origin or religion is a fairly big problem in their local area'. This may overlook the experiences of an ethnic minority in an otherwise culturally homogenous area, where a majority of people don't see any problems. In Blyth, the BME (Black and minority ethnic) population is very small but the Bangladeshi focus group told me of some fairly serious instances of intimidation and criminal victimisation members had suffered recently.

The AC set includes no indicators that measure adequacy of income in relation to needs, income security or control of debt which showed as major concerns in the Blyth Valley data. The AC indicators, although they focus on benefit claims and residents of deprived areas, tell us little about the disparity of incomes in the borough and the gap between wealthy and poor. The AC set measures certain aspects of the labour market (job density/employment levels) but offers no way of evaluating the quality of jobs. The same is the case for its education indicators. It would be hard to get a clear picture from these indicators of the quality of activity generally, of who cannot and/or does not access educational and leisure facilities and why; neither do they include anything on mental health. This indicator set would be incapable of representing the social isolation, boredom and depression that were evident in the Blyth Valley data. It has an indicator to measure teenage pregnancy, which was not a concern voiced by any of the public, although a few policymakers mentioned it.

Where the AC set was relatively strongly correlated to the Blyth Valley account was in the environmental quality or physical surroundings theme. This is unsurprising since its focus is on local 'liveability' and the perceived responsibilities of local authorities. However, there are no indicators which adequately reflect the importance of the experience of the environment to people's wellbeing (as discussed above). For instance, it doesn't have any indicators for tranquillity or access to allotments or gardens. Nor does it measure housing security; its focus is on the quantity, quality and affordability of dwellings which of course can be related to levels of housing security but given the emphasis placed on the latter in the Blyth Valley data, I would have preferred a direct measure.

Subjective wellbeing and sustainability

Neither Nussbaum's list nor the AC set represent a number of wider issues which occurred in the Blyth Valley account; these included environmental change, global and national inequalities, political and economic systems, social norms and values. Some people mentioned that 'peace in the world', reduction of global poverty and using fair-trade goods were important to them:

> You have no idea how it took a year to get fair-trade coffee in our meeting rooms, it took a year, and this is an organisation with freedom over our budget and I sent minutes and I sent notes to senior management to persuade them and then there was complaints that the coffee wasn't as nice, you've no idea, why is it so bloody hard?
>
> (Policymaker)

> As a member of the Labour party we've always been committed to internationalism, whether you like it or not we're interdependent globally. I've always believed that, I'm trying, in my own little way, but it's better than nothing, trying to get people to take on board again that we have a responsibility. International socialism is about people self-reflecting and saying hang on this is not fair, this is totally out of order.
>
> (Policymaker)

One person talked about the extra fear in society post 9/11 and the resulting restrictions on civil liberties. Others talked about capitalism, economic systems and the power of the private sector: 'We are run by a capitalist agenda; we are about economic issues' (Policymaker); 'Capitalism is a force. It's a force that makes people go out to work to earn the money to buy the goods that will make them happy, but don't' (Policymaker). Some policymakers expressed the view that what they were able to do at a local level to effect change was very limited due to the power of larger economic systems. For example, one person described how the area partnership had a very limited budget to promote a five-a-day healthy eating campaign but felt they were ineffective against competition from international fast-food giants who have a huge budget for advertising and promotion. That person felt they would like to see an indicator which reflected this sort of issue.

Many policymakers, including many of those without professional responsibility for environmental policy, had environmental concerns and values, despite the marginalisation of environmentalism in the council:

> I mean with global warming and the threat of that, particularly to third world countries, can be astronomical. And we ought to be discussing it now and we ought to be discussing it at a local level because we can do so much locally. And it ought to be embedded within our whole culture about sustainable resources, about our own population...that's going to have astronomical effects on long-term resources and the long-term health of the planet. But it's not going to happen by people thinking, 'Oh well, the government will sort that out.' We ought to be concerned here and now. We ought to be concerned about the amount of waste that we're producing in our country, in Europe and the States.
>
> (Policymaker)

Although only a few members of the public explicitly expressed concerns about the global environment and global justice,[2] many more were concerned about increased materialism, individualism, debt and the need to live more simply – issues which obviously resonate with sustainable resource use, responsible consumption, and global justice. Rising materialism was frequently linked to a decline in social values and, interestingly, social values of care and living more simply were intertwined. In addition, local environmental quality and amenity were often strongly associated with the idea of living more simply, being linked to childhood memories of liberty and play. The importance of personal qualities and positive emotions were strongly linked in the Blyth Valley study to social values and freedom. This maps well onto evidence emerging from subjective wellbeing research which shows a much more complex relationship between income and wellbeing, and the importance of social relations and psychological needs. As outlined in Chapter 2, new discourses of wellbeing promote a link between happiness and a dematerialised lifestyle, often focusing on 'making a contribution to the community'. Although the idea of community was important to many respondents, for many others it was not a focus for their views on wellbeing; in fact, for some the idea of a 'tight-knit community' had negative connotations. A wider, richer idea, an idea of 'social values' which is inextricably linked with people's lived experience of themselves, each other and the environment was strongly mobilised in the Blyth Valley account and it is perhaps this, rather than a (sometimes shallow) focus on community cohesion or community participation, which could open the door for more legitimate discourses of social quality and common good, in which environmental and global justice issues may receive a more sympathetic hearing.

In this sense, both Nussbaum's list and the AC set are lacking. Nussbaum's focus on capabilities and her desire to avoid prescription means she offers little guidance about social values in relation to the material and economic world. The AC set is founded on community relations, and aims to prevent anti-social behaviour rather than promote specific social values. The two are clearly linked but according to the AC set it is possible to be a model citizen (no crime, vandalism, graffiti, drunken rowdiness, pleasant to and tolerant of your neighbours), while lacking social values of empathy, hospitality, care and compassion. In these themes, the evidence on subjective wellbeing and its focus on psychological needs and resilience, the quality of social relationships, 'intrinsic' values and less materialism resonates better than Nussbaum or the AC with the Blyth Valley material. I felt that a combination of Nussbaum's capabilities and the subjective wellbeing evidence fitted with the Blyth Valley account. Interestingly, emerging evidence, using Nussbaum's set of capabilities and statistical data on SWB, finds that from a substantive point of view each of the ten factors of capability, if present, largely affect SWB in a positive way (Anand *et al.* 2009). In addition, I argue that they are complementary. The universalism of the capabilities approach provides the necessary normative power and protection against injustice, a view strongly mobilised in Blyth Valley, while SWB evidence may provide evidence to support local views regarding those aspects discussed above which Nussbaum does not account for. However, the gathering of detailed data on local wellbeing, and understanding wellbeing as a lived experience which is dynamic can provide contextual and cultural specificity to guard against the imposition of inappropriate blanket policies.

Summary

The Blyth Valley work shows the value of building an account of wellbeing based on qualitative inquiry; it resulted in a richer, fuller picture than Nussbaum's list or the AC set. Nevertheless, the picture is infinitely more complex and the resultant account needs further refinement before it can be applied for political purposes. The Blyth Valley account includes resources, capabilities, needs and happiness; for philosophical purists this is a disaster. However, for policymaking, conceptual clarity is probably more important than philosophical purity. Local authorities cannot always provide directly for wellbeing; they can sometimes provide resources, address needs or enable capabilities. A concept may work not because it is philosophically consistent but because it is politically legitimate, its elements are internally coherent, it gives clear policy guidance and because it has meaning for local people.

Most aspects in the Blyth Valley account can be found in a general sense within Nussbaum's list, although a number (e.g. resources and subjective wellbeing) are missing for conceptual reasons. However, her focus is on the individual and the social institutions that support capabilities, rather than the interconnected neighbourhood. The AC set, despite its focus on local environmental quality, liveability and community relations, is weak in highlighting the importance of personal and social relationships, and does not deal well with the importance of environmental experience. So the capabilities approach offers a useful normative basis for wellbeing policy; local people prize justice and freedom, it has resonance with SWB evidence, can be modified to suit local needs yet provides a universal account of social justice. It can underpin a local conception of wellbeing which local views could enhance. In terms of human wellbeing, the social justice account of quality of life based on the capabilities approach is appealing.

However, there are a number of issues which perhaps the account does not address and which need extensive debate. The prime message from this exercise is that the things which are difficult and contentious are among the most important, and to conceive of wellbeing only in terms that are easy to measure is to side-step many of the deeper issues that affect us all. Neither Nussbaum's list nor the AC indicators could deal with individualism, materialism, ecological sustainability, economic systems or social value. Some aspects of wellbeing are difficult to affect because they are related to structural issues like global economic forces which individuals, communities, local authorities or even national governments have little control over. As such, without radical change, some of these quality of life and wellbeing issues will remain unmanageable. What is needed is a fuller debate about who should be responsible for particular issues and the role of local governance in that.

9   Moving towards measurement?

I have illustrated how this indicator project was set within a challenging political and discursive context which exacerbated tensions in interpersonal relationships in the indicator team. The initial aim of creating indicators through a collaborative process with the local strategic partnership faltered, although consultation and qualitative research was conducted to determine a meaningful basis for conceptualising wellbeing. This work resulted in a Blyth Valley 'wellbeing framework'. Many questions remained unanswered. How were inherent tensions to be addressed? What should determine the choice of indicators? Who and what are the indicators for? How do they relate to sustainable development? How would they influence policy decisions? The answers to these questions would decide the design, focus and resourcing of those indicators. Funding to invest in data collection for new indicators would have to justified on the basis of their perceived importance, which in turn would be linked to conceptions of wellbeing, and of what the council were trying to achieve, for whom and in what way.

Although people talked freely about what they thought was important for wellbeing, discussions about measurement and developing indicators posed a substantial challenge for policymakers and residents because of the tensions between different perceptions. Typical reactions from the public included: 'You can't really measure it, it's so individual'; 'You cannot measure quality of life, it's not an easy question'. This dilemma was also frequently expressed in the interviews with policymakers, who struggled with the problem of capturing this complexity and diversity in a single set of indicators:

> Is it possible to measure quality of life given that individuals are all different and how do you go about that anyway? What does it encompass?

> To me perhaps the problem is you either tackle it, and you probably can't, from a very high level, the general question of, 'People, do you think you've got good quality of life?' Yes. And maybe as an argument we say that's the best indicator you can have because people's quality of life is a personal thing. It's this for me. It might be that for you. It might be that for the next person. Ask a simple question, or be very scientific, and drill it down into hundreds of indicators…how do you measure it?

Box 9.1 Excerpt of focus group with 14- to 16-year-old pupils from Blyth Community College

The problems with measurement

Researcher:	Can you measure it [wellbeing]?
Student 1:	No, everyone wants different things.
Student 2:	It would be a massive list...
Researcher:	What about some of these standard measurements like percentage of people in employment?
Student 1:	No, because most people hate their jobs, you should ask how happy people are in their job...
Researcher:	What about teenage pregnancy rates?
Student 1:	That just measures how stupid you are.
Student 2:	Yeah, and all the support is around that.
Student 3:	A lot of the support is all just sexually-based, contraception and that ...
Researcher:	What about this, the percentage of pupils attaining five GCSEs?
Student 2:	People can be successful but that doesn't mean they are happy.
Student 1:	They measure how well people are doing but not how happy they are.
Student 3:	Can you just ask people how happy they are?
Student 1:	No, they lie. Like when people ask 'how are you?'
Student 2:	No, because they could be showing happy on the outside but on the inside they are really not happy.
Student 4:	Also if you've had a really bad day it's going to affect what you say.

These discussions also highlighted the inadequacies of existing indicators or targets, the things they did not measure and the ways in which they affected policy decisions and resulting programmes and initiatives. I asked a focus group of female pupils (aged 14–16) from Blyth Community College, to consider wellbeing measurement by looking at some of the standard indicators. They reflected on some of the key difficulties (see Box 9.1). Some of these concerns were echoed by policymakers, many of whom desired better data but who also had mixed feelings about the nature of indicators and about the role of the local authority in determining which aspects of wellbeing to measure.

Role of the local authority

As discussed in earlier chapters, many people felt that the role of the authority was to help provide quality of life context (for example, through housing, health, education and local environment). Policymakers often, for good reason, focused on tackling deprivation, which they discussed in terms of 'getting the basics right' or providing for 'basic needs'. Intertwined in these discourses was concern not only with structural problems and inequalities (for example, in housing, education and employment) but also with what many described as 'cultural' factors which included low aspirations and individual attitudes (for example, to work, health and education). There was often a marked and/or explicitly articulated difference between how policy actors viewed their own wellbeing, and how they conceptualised it more generally in terms of their work.

> The thing that hits me the most is when sometimes you talk to people about educational attainment, and parents and kids think that the education…is great. Hey, there's not a problem. There's an issue there about aspiration. Now, we shouldn't be forcing different aspirations on people but I think we should be opening up, opening the curtains to a window which has got opportunities, which allows aspirations to go wider. So if that's what we're trying to achieve, that should be it. We shouldn't be trying to force changes on people. But sometimes people think things are okay because they don't know any different. And there's a huge issue there about aspirations.

> I think it's really the basic stuff, you know, what's that thing like? [tried to remember Maslow's hierarchy of needs] It's like what you need is shelter, food, education, all those really basic things, I think are the quality of life issues [for some residents]. I don't think they ever get the opportunity to think about the stuff that I would think about my quality of life, like being able to walk my dog, having a nice neighbourhood to live in, or good neighbours, or being peaceful and at home. I think their quality of life measures [are] about surviving…I think, and it's difficult to stand in their shoes, but I think it's got to be, you know, the basic stuff.

Some contrasted their own post-materialist values with the values that they projected onto residents, raising the difficulty of integrating different perspectives in the development of indicators:

> A lot of people…wouldn't have the first clue as to what you mean by quality of life indicators but some people might think that because they've got a satellite dish, that's a quality of life indicator…I don't strive to own a big flashy house and I don't, because that's not what's important to me. So maybe there is quite a difference, yeah. I do understand what people want. They're very materialistic, a lot of people are materialistic and I'm not.

Some policymakers were reflective or critical about how discourses of quality of life were reflected in a top-down approach that held various assumptions and prescriptions. One policymaker warned against 'a sort of quality of life fascism'; this reflected the concerns of a number of others who questioned whose values quality of life indicators represent:

> I'm not certain that there's starting to be a set of values, liberal left middle class values that define what quality of life is and everybody must fit into that. And I, you know, it just raises some questions in the mind as to who are we doing it for and why?

> 'How do you feel you fit in with your community?' I have personal difficulties with that as an indicator because I'm, you know, having joined the middle classes because of my income and where I live, I don't feel part of my community 'cos I'm out at work for the best part of the day. And I go back and my community is more likely to be my friendship groups and my family, it's not likely to be the lady next door or, you know, the WI that meets once a week. But yet we expect deprived communities to say, 'I feel part of my community' [or] 'I talk to my neighbours'. I think it's a bit of a double standard.

> Who are we to say to the person in Blyth…whose kids are probably not going to attain average educational attainment of key stages, whose family have got a history of medical problems, whose housing is less than satisfactory but who say our quality of life is actually really, really good…we're a really happy family. As opposed to somebody…who's struggling to keep up their mortgage payments or pay the charges to keep the horse in the stables and who has major problems with their marriage, their quality of life could be crap…How on earth can you build up indicators that try and put those people on a level playing field to say quality of life?

These discussions were often circular and returned to the point where policymakers reiterated that quality of life was different for everyone and that one set of indicators could not meaningfully represent different needs:

> We have mixed communities these days, different cultures and backgrounds, social interest groupings, disability, race, sexual orientation you know all that sort of stuff, economic disadvantage is in there as well. And if you asked what quality of life meant to their lives, you know, you might have a different way of looking at it or approaching it, instead of saying we can just have one set of indicators that applies to all, well that doesn't measure anything does it?

I have tried to represent here the complexity and circular nature of these discussions. My qualitative research method aimed to avoid paternalism as far as possible, by seeking to produce a broadly inclusive account with wide resonance, that allowed policymakers and residents to reflect on the most important things in life to them. The wellbeing framework had less of an inventory feel than some other accounts, being developed from complex discussions of inter-related factors, and it allowed a more nuanced view of wellbeing than the council's existing discourses of development. However, it did not have the power to resolve inherent tensions in values.

The role of indicators in policymaking

The difficulty in generating interest in indicators was not only a result of the considerable conceptual and normative difficulties described above. It was also a product of the widespread cynicism about the role of indicators, and about evidence and targets in policymaking. One aspect that policymakers discussed a great deal was how centrally-set targets with funding attached geared policy in particular ways, either to address specific issues determined centrally, or to direct it to areas where there was a better chance of meeting performance targets:

> Often they're influenced by funding, like what funding there is out there that we can tap into. So, for instance, in southeast Northumberland we've got neighbourhood renewal funding, in one respect, plus neighbourhood renewal funding in Blyth and that's very much directed, very focused on specific areas of activity.

> It's a bit like the index of multiple deprivation. We use it, we use it as a tool, we use it for a tool to lever money. But other than that does anybody really do anything with it? Do you know what I mean? You don't actually do anything with it. It, I just, I don't mean what you're doing is worthless, I don't want it to sound like that...I just, I don't know how it would be used as a council, by the council.

One policymaker described his frustration at a regional policy decision where a pot of money to encourage entrepreneurial skills in young people was targeted at 'good' schools where there was better chance of meeting output targets:

> If we've got, whatever it was, £300,000, I can't remember if it was more than that, don't plough it all into the areas that already have. Start investing in the areas that haven't. ...The outcomes from the very good schools are probably going to be much better than the other [poor] schools. But somewhere along the line you've just got to bite the bullet and say would that still be value for money.

This funding-led, target-driven culture was linked by some to central government's focus on evidence-based practice:

> That would be one criticism I would make, that this government particularly has used evidence-based practice to say, 'A plus B we will get C, and if you don't get C you've failed.' And people aren't like that; people don't fit into

those categories and boxes. I think there's good practice, we can stop making the same mistakes…but I do think we don't pay enough attention to failures because that's not, that's not good press for anybody, to point out failures…I think if you are doing evidence-based practice then it's got to be on what doesn't work and we don't get the honesty for it to be robust enough to use that. Will I go out and sell what's not working here? No…The whole system doesn't encourage me to be honest.

Often the timelines in which central government required targets to be met were unrealistic, due to the complicated nature of issues and the difficulty in measuring them:

There is a health, mental health promotion strategy for Northumberland which is divided into localities and what we're doing in each area, so there are things happening in Blyth Valley that are trying to address mental wellbeing, as well as the other physical ones. But it's very difficult to be able to say after two or three years what might have made a difference or if a difference is being made.

This led, in some people's opinion, to a focus on specific 'deliverable' targets which were important but risked ignoring wider issues:

Government targets don't help. I think targets are right but they still don't help because people are working towards what their targets are and the mindset and the focus is to deliver that and not necessarily thinking about the bigger picture.

If we take the elderly, the only measure in the Local Area Agreement at the moment for the elderly is what the NHS happily call the Frequent Flyers, you know, the elderly will trip over a lot…that basically suggests that the only issue we need to address for that group of people is that they fall over and injure themselves.

These discussions also related to the need to consider local context and different types of evidence. Some felt that no matter how many indicators you have they could only give a partial picture and depending on what is measured, may present a distorted view of problems for certain groups. The idea of a need to look at the 'bigger picture' not only focused on having a wide view of all the different aspects which interacted to produce wellbeing, but also methods of how you can gain that bigger picture, to avoid a shallow tick-box approach:

I've got a raft of information, MORI polls, surveys that we've commissioned and I don't use them as anything other than a tool, it's one tool in an armoury of things and it's one snapshot that says to me there's something happening and if I just use that one snapshot, it's like taking a picture of a smiling family and saying that family's happy.

Integrating the qualitative issues into a quantitative measurement system is so vital to really make a difference, you know, it's easy to set up training courses and get a hundred people on it and tick your box but it's the stuff like making sure you get the right people on that training course.

In the minds of some respondents, this reinforces the value of subjective indicators and qualitative research in complementing the bald statistics of objective indicators, to get at some of the hard to measure aspects of life which are important. Yet whilst recognising this value they acknowledge the difficulty of data collection and interpretation. In addition, those who were working closely with people often felt they knew the qualitative issues, through their experience, and resources would be better directed at delivery and that indicators were a distraction:

My first and foremost basic concern is that the more indicators you have, the less people will make an effort…things that make a difference are qualitative and not quantitative and therefore it becomes so much time in the process of measuring them that you could spend actually doing it.

There are some areas also where we know that it is a good idea to do stuff, but we haven't got any way of qualitatively or quantitatively justifying it.

I mean I think the things that are called 'softer measures' (which I don't actually think are)…this is different and you have to adopt different approaches. I think softer measures are perfectly hard measures [hard to measure]… So in taking those softer measures I think they're really useful measures, but we have to find ways of interpreting them well.

In summary, thinking about measurement was very challenging for people. This was due to conceptual difficulties of how to define quality of life or wellbeing, whose values should determine it and what should be the role of the local authority. More than this though, was a striking cynicism and reservation about targets and indicators and their role in the culture of policymaking and practice. It was also clear that developing indicators was a researcher's job. Comments of 'rather you than me' or 'good luck!' were frustratingly common. As discussed in Chapter 7, I argued that indicators could not be made by me without a collaborative discussion about some of these issues, that they were a political process and reflected priorities and trade-offs.

Integrating sustainability

As seen in Chapters 6 and 7, within Blyth Valley there were many conflicting perspectives on the relationship of wellbeing to sustainability and how to bring the two areas of work together. Understandings of sustainability present a complex picture. One way of understanding this controversy is to distinguish between 'weak' and 'strong' conceptions of sustainability as set out in Chapter 3. However, these are over-simplistic contrasts, divorced from the complexity of real-life situations. A person rarely exhibits only one standpoint which is unshifting for every context and situation and the data around sustainability was complex.

For example, a senior policymaker whose job was focused on economic regeneration, described to me in his interview how at home he was continually nagging his teenage children to recycle and was frustrated at their lack of concern about climate change. He said he felt worried about their quality of life in the future, due to looming environmental problems. However, this person was constantly criticised by other officers in the council for aggressively pushing an economic agenda and side-stepping environmental issues. He was very comfortable using the language of regeneration discourses and at times was disparaging of 'environmentalists'. His views on sustainable development were complex, he felt pressured to deliver physical and economic regeneration in the borough and was influenced by institutional discourses, even though he had deep personal concerns regarding environmental change. However, his interview illustrated a common discourse, that environmental concerns were a luxury of the middle classes and therefore, for policymakers, almost a private concern which should not take priority over addressing the 'basic needs' of deprived communities. This was heavily influenced, sometimes explicitly, by Maslow's hierarchy of needs where environmental concerns were higher on the ladder and would only be addressed when 'basic needs' are sorted. For many environmentalists, this ignored the interconnectedness of environmental issues and inequality. In terms of measurement, the difficulty was expressed as a dilemma between different paths or of 'balancing':

> It's easy to see particularly in relatively deprived communities that quality of life is a very immediate issue and one which is – is pretty much independent of global issues. I find it personally quite hard to reconcile the two, you know you can – you can create a sustainable environment which also, you know, scores highly in the quality of life indicators, but it tends to be very expensive. So I think that there is – there's a balance to be struck between the two and so yes – certainly if you look at the kind of indicators which are used – the indices of multiple deprivation, most of them are not the intergenerational sustainability are they? But they are quality of life indicators.

Although individual officers expressed environmental views in their interviews, in an institutional sense, SD was not well understood or promoted in BVBC, apart from a few officers who tended to be marginalised as discussed in Chapter 6. Therefore, this limited the space for collaborative discussions of how wellbeing maps onto sustainability to find legitimacy. These discussions were not well progressed even though there was plenty of scope in the wellbeing data to highlight the importance of sustainability issues. I argued therefore that work around developing indicators should be separated into defining wellbeing first and then looking at the trade-offs and tensions between that and sustainability to offer more transparent discussions.

Changing the discourse

This research found that BVBC wanted to develop indicators based on participatory processes but this meant different things to different people, and the role of indicators was seen differently by different people. This was limited in practice to consultation and qualitative data collection from which key policymakers then expected one or two people to produce a set of indicators. However, this would miss out on important aspects of participation and collaboration in the decision-making process related to the discussions above, and would not address the deep concerns people had about indicators, therefore lowering the chance of indicators to be valued or effective. For this reason, I resisted the pressure to produce a set of indicators from the data. In any case, as I try to show, these issues are so complicated that one person alone, no matter how skilled, could not create a robust and democratic product. Some policymakers urged me to create a set of 'Aunt Sallys' for discussion, and whilst this argument has some merit, I was aware that this had happened twice in the past and the indicators were simply 'signed off' with no further discussion. I was, therefore, reluctant to do this, having argued for so long for a more robust and critical approach. In any case, it would not have been possible for me to take the work further without a greater input of resources and a much clearer steer from BVBC.

The Blyth Valley wellbeing framework was presented to the LSP in April 2007. Alongside the wellbeing data I discussed the issues with indicators outlined above and the need to discuss how sustainable development issues could be addressed. This meeting was the fullest meeting of the LSP Board that I had attended. I was very pleased to see so many people there, including the whole indicator team. The presentation was received with interest and a healthy discussion ensued. Unfortunately, despite agreement for a period of debate and a 'stakeholder event' or conference, neither of these things happened. This was partly because of the Local Government Review, and partly because, as already described, there were institutional barriers to creating an effective forum for collective deliberation.

The initial objectives of the indicator project were not realised and 'concrete' outputs (a set of indicators) were not produced. However, a deeper engagement with and understanding of the concept of wellbeing was developed by key officers. Action was traded for argument. Deliverables were traded for deliberation. This meant frustrations, tensions and slow progress. Participants involved, including myself, were challenged to look at their assumptions and practices. In the words of the strategic planning manager 'we didn't get what we expected, but we got something deeper and more challenging'.

The following statement was included in the 'Strategic Objectives for Blyth Valley 2008–2025' document (BVBC 2007), written by the strategic planning manager under the supervision of the executive director:

Feedback from partners through the local strategic partnership (LSP) recommended that we develop more qualitative measures using surveys and focus groups. We have also helped to develop a wellbeing framework giving some draft thoughts about wellbeing indicators. We presented an initial report to the LSP describing the background and reasons for adopting this framework. The framework will enable the LSP to develop its own set of measures reflecting local experience and ambition. In some areas, particularly economic development, wellbeing measures tend to conflict with more traditional regeneration measures and in some cases, with sustainability measures. Wellbeing measures tend to centre on the individual while much of our work currently focuses on communities. As this work is still in its early stages, we have suggested some indicative wellbeing measures for consideration. Although we currently identify these separately, we will need to integrate them into our future work. Where measures conflict we will need to identify the most appropriate way forward.

(BVBC 2007)

I am not suggesting that this example represents a complete change of discursive positioning on behalf of the council. However, it does represent a definite *shift* of the key policy actors involved in the indicator project away from the dominant discourses and towards greater reflexivity and acceptance of conflict and complexity. This was a significant change from previous policy statements, which tended to promote a more simplistic view of wellbeing based on economic regeneration. There was also some recognition towards the end of the project, by those previously advocating rational evidence-driven approaches, that the process was much more complex and all encompassing than they had originally thought and needed political engagement at the highest levels. At the LSP board meeting where I presented my final thoughts on the indicator work, the executive director expressed the view that the work was 'extremely influential' and that the council would perhaps need a fundamental repositioning in the way it approached development.

The sustainability officer, in a final report evaluating the indicator project wrote:

A substantial amount of time was, and still is, needed to think through what these concepts mean (quality of life, sustainability, wellbeing etc), and how they related to the vision for the LSP, and how it can be achieved. On top of that, looking closely at how indicators work, and the need for a deep understanding of their potential and limitations is also necessary.

I personally have learned quite a lot from this (project), in terms of substantive knowledge on the quality of life/wellbeing/sustainability debate, and indicator development in general. I have also learned about myself, about how I react in certain situations. I have reflected upon my role as a manager, and how I operate professionally when there are difficult decisions to be made, and conflict and discord in a working environment.

Last word

From a Blyth Valley resident:

> I think there will be as many ideas of quality of life as there are people, and the achievement of what you want can be helped by the authority, but not provided.
>
> Good luck, it's a toughy.

10 Towards sustainability

Developing a common sense of wellbeing

This book has reflected on wellbeing measurement and its relationship to sustainability, participation and power, evidence and knowledge in policymaking. It related my experience of being part of a research project to help develop a measure of quality of life for political purposes, specifically for use in local governance in the Northeast of England, and to understand its value in promoting wellbeing, sustainability and local democracy. Since that project began in 2004, wellbeing has risen on the political agenda accompanied by a serious endeavour to create a set of wellbeing measurements for policy and this in turn influenced the work at local level. However, it seems clear that no matter how intelligent the indicator, or how intense the search for it, the complexity and nature of wellbeing cannot be fully reflected by such measures, though statisticians and economists will surely try their best. Although statistical wellbeing measures are important for many reasons, not least the debate they engender, they can only ever play a partial role in policy discussions about development. Practical judgement, contextual experience and empathetic values must also be given a legitimate place in democratic policy debates that are sensitive to issues of power. However, this means building a culture where wellbeing is not only a condition of individuals or neighbourhood communities but also of social and political institutions. This poses deep challenges for the concept of wellbeing and its measurement.

Defining wellbeing and sustainability

As outlined at the start of this book, previous studies exploring the development of local quality of life indicators reveal a dilemma. In trying to engage local people using terms such as quality of life and wellbeing, discussions tend to reflect local 'liveability' concerns and miss the global, intergenerational and ecological dimensions which underpin sustainable development. The synergy and tensions between wellbeing and sustainable development will, of course, depend on how these things are constructed. In liberal democracies, for the past few decades, the dominant construct of wellbeing has been an individual one, reflecting a complex meshing of ethical ideals of justice and freedom with neoliberal marketised views of the self. The concept of sustainable development therefore has been largely framed within ecological modernisation narratives which are underpinned by a 'win-win' scenario of environmentalism and economic growth for better quality of life.

The case study described how key policymakers in Blyth Valley Borough Council promoted economic regeneration as a dominant strategy. This focused on attracting investment, promoting physical regeneration, improving the socio-economic profile of the area and changing an 'inward looking' culture. This was underpinned by a strong social justice ethic of empowering local communities and reducing inequalities, where policy gaze was directed to addressing geographical 'pockets' of multiple disadvantage. However, quality of life was a vague under-developed concept, often understood as the inverse of deprivation and policies were justified by reliance on IMD type statistics. Although policymakers had strong social justice values, these were often mobilised in discourses which focused on basic needs and created a simplified and stripped-back account of what wellbeing is for such communities. Although serious inequalities and deprivation existed for sure, these discourses made assumptions about poor people's aspirations, needs and preferences, assumptions which policymakers themselves, when questioned, seemed uncomfortable with. As Woodward (1996: 58) argues, 'deprivation' has become associated with a 'rather monolithic set of criteria relating to material conditions'. While understandable, this can exclude a range of experiences important to people. In addition, ill-defined notions of wellbeing can make it difficult to explicitly discuss theories of cause and effect and what the solutions might be (Ganesh and McAllum 2010; Seedhouse 2001). Many policymakers could not articulate a convincing link between economic growth agendas and reducing inequality, and relied on additional measures like community development and neighbourhood management initiatives to create change in disadvantaged areas.

Environmental concerns were legitimised if they were judged to have socio-economic benefits, such as alleviating fuel poverty through the provision of renewable energy or the creation of a pleasant environment to attract business. This reflected central narratives of eco-modernisation which underpinned the 'sustainable communities' agenda and a focus on localism and place-making. Other environmental concerns like air quality, sea level rise, climate change or biodiversity, which tended to conflict with regeneration narratives, found little purchase. Environmental protection regulations were often viewed as hurdles to be jumped. Although many policymakers held environmental values, these were often seen as 'private' values for middle class people and judged as largely irrelevant for disadvantaged people. More outspoken environmentalists were characterised as 'putting the environment before the people'. Similarly global justice values were held by many policymakers but they struggled to find a place to articulate these within institutionally legitimised narratives of 'let's get our own problems sorted out, then we can save the world'.

In this study, the concepts of quality of life and wellbeing were explored in detail, not as synonyms for sustainable development, but as separate concepts. In consulting with local people the project did not seek to educate 'communities' about SD, although this was perhaps an original intention. Nor did it seek to be representative of 'community' views, recognising the inherent complexity and contested nature of both wellbeing and sustainability. The indicator project explored local understandings of wellbeing, quality of life and happiness without framing wellbeing in terms of local residential experience. Although there was much 'overlapping consensus', people put different weights on different aspects of wellbeing and had different theories about how to create it. However, a general account of wellbeing was produced which reflected a wide range of views and values from local residents and policymakers. This 'local' account of wellbeing found strong resonance with both the capability approach (Nussbaum's list) and with evidence on subjective wellbeing, for example that social relationships, giving, living simply and general levels of friendliness, care and trust in society are important. The work also reflected a number of wider concerns local people had about how individualism and materialism, social and environmental values, political and economic systems, justice and fairness affect wellbeing. Not only do these issues create a challenge in terms of measurement, they necessitate a clearer debate about what sort of values should drive development. As such, a deeper and wider discussion of what wellbeing means in the context of everyday life, related to a discussion about underlying values, may have more resonance with sustainable development ideals.

Democratic processes – participation and power

Local participation in Britain during the time of the study was heavily prescribed by central agendas, one of which, ironically, was community participation. Central strategic policies of promoting 'sustainable communities' during the New Labour years provided the conceptual and evaluatory basis for increased localism and the development of local quality of life measurement. While containing considerable normative content shaped by ideals of social capital and community participation, these strategies were influenced by the same modernisation discourses as discussed above. Certain forms of participation were encouraged and legitimised which created tendencies to focus on consensus, not through an engagement with conflict, but through marginalising dissent in various ways. In addition, a focus on empowering communities, whilst important, lacked critical content and limited the scope for looking at wider structural issues as cause and solution of local problems, leading to an effect described by Amin (2005) as 'localizing the social'.

In the case study, I illustrated that despite the commitment of BVBC to engage and empower communities this was inhibited in practice by a lack of critical engagement with power relations and a predominantly rational view of policymaking. Much thought and many resources were given to the techniques and procedures of consultation, the timings, setting up a database of responses, ensuring all residents were consulted and within a particular timescale. It was assumed that all this information, some of it conflicting, could be somehow processed and fed into policy. However, it was very unclear how and in what ways this was actually done. A much greater emphasis was placed on actual consultation events than on how decision making was affected, why certain values were undermined and others promoted.

The lack of an effective forum within which to discuss these issues transparently was continually frustrating to the project of defining wellbeing for public policy. At a policymaker level, there was early interest in discussing this amongst the LSP members but this interest waned as the LSP were seen as less and less effective; as people were overwhelmed with new initiatives; and as policymakers indicated that there was no need for a forum because community-led development of indicators was something a researcher could do through some sort of qualitative/quantitative analysis of the community's priorities. Many policymakers expressed frustration at 'talking shops' and saw the development of indicators as a technical rather than deliberative exercise. Respecting and managing conflict is key to collective deliberation, however, in BVBC and the wider partnership there was a lack of institutional capacity to handle the various conflicts that arose and to view them as an important part of the democratic decision making process. This means that there was limited capacity for a truly discursive and democratic space where values, definitions and the development of indicators could be negotiated transparently.

A local qualitatively developed account of wellbeing was produced through research with local policymakers and residents. Although the study involved local people through interviews, focus groups and consultation, these methods were largely 'extractive'. They were appropriate methods for gaining a wide overview of what people felt was important for wellbeing but this must be developed further through debate and a deep questioning about which values should underpin a local account of wellbeing and what effect this would have for whom. How to create measurement of such an account will necessarily involve tackling tensions, prioritising some aspects of wellbeing over others and reducing wellbeing complexity to manageable chunks. Indicators must be produced through discussion and debate as their production is essentially a political not an academic process. However, finding an appropriate forum for this debate and the resources and interest for it was very difficult. In many policymakers' eyes the community was 'out there' and they often focused on disadvantaged areas. Whilst some policymakers idealised and legitimised the community view, consultation responses which conflicted with their construct of the community were often overlooked. There was little space for a more complex, realistic appraisal of a variety of views.

Potential for influence on policy

Indicators are often seen as objective tools, developed outside the policymaking process, which can be used by policymakers in a step-wise way to inform rational, evidence-based policy. Where participation happens it often focuses on consultation rather than debate, producing an 'inventory' approach to indicators, where everyone's interests are represented but providing no guidance regarding inherent tensions. Concrete change in policy is often a key criterion for indicator success. Indicators often fail this test as research has revealed little substantive change. This is unsurprising as processes which are mobilised in the development of indicators are often influenced by the same discourses which they seek to change, inventory lists provide no clear steer for policy decisions, more and more data does not help and policymaking does not follow in a step-wise way from evidence gathering. However, if power and discourse were widely acknowledged in institutional discourses as key factors in policymaking processes, indicator projects would have a very different focus, design and evaluation criteria.

During my time at BVBC, there were many practical obstacles to the indicator project; a wide range of structural and external factors, such as the local government review and the introduction of Local Area Agreements, affected the work. The rapidly shifting policy landscape created an uncertain ground for this non-mandatory indicator project as it was difficult to make arguments for it in comparison with the latest government initiative. Policymakers found it hard to create the time, resources and space for reflexive discussion, debate and deliberation. However, the project was also hampered by a predominantly rational view of policymaking by senior management where indicator work was expected to happen largely outside the policymaking process. These views were also bound up with assumptions and understandings of 'hard' and 'soft' data which made it harder to argue for the validity of qualitative research or experiential judgement.

Despite rhetoric from central government about the necessity for quality of life indicators to be locally derived, the indicators which were controlled at central government severely limited the room for manoeuvre at BVBC. Although policymakers viewed such indicators as having limited value and were cynical about how they created a tick-box culture, they were driven to gear their work around statutory performance indicators or the new LAA indicators. This cynicism affected the appetite of policymakers to consider producing a further set of local indicators and they struggled to see how this would have any influence. Comparative statistics are vital to ensure standards are maintained across the country, but in order to make comparisons, a set list of indicators has to be applied to all local authorities, and it is this list which authorities will continue to use to steer their priorities.

This work tried to create a meaningful account of local wellbeing to add greater depth to policy discussions. In terms of promoting wellbeing, developing these understandings through such discussions and processes can promote a greater awareness amongst policymakers about the complexity of wellbeing which becomes part of their everyday practice. It can help them to re-evaluate what wellbeing means and hence what policy is for. Discursive and experiential processes help us to internalise such understandings which then become a part of our everyday practice. There was a sense in Blyth Valley that more evidence was good and that policy could not proceed without detailed 'baseline data'. Whilst this sort of evidence is important, the need for it is often illusory as policymakers will only tend to use the data which supports their priorities. Waiting for the perfect set of data is an illusion and can actually paralyse decision making and lead to a lack of conviction and confidence. Internalising a complex understanding of wellbeing means that policymakers may rely less on large amounts of data or indicators and more on political processes. At a large Northeast conference to look at SWB in 2007, the majority response from policymakers was 'very interesting, but what do we do with it?' However, as Des Gasper (2007a: 58) puts it (with reference to Martha Nussbaum) perhaps not everything that is important needs to be measured, or can be measured, instead an alternative is to ensure the 'institutionalisation of the concepts of wellbeing' by producing 'rich qualitative description' to promote 'greater understanding and reorientation of policymakers'.

The officers involved in the indicator project went through a process of conflictual discussion and reflection about their policies and practice on wellbeing, quality of life and sustainability. There was therefore a much greater awareness of the complexity of these issues and how to implement them at a local level at the end of the project. This demonstrates an increased capacity at senior and middle management level to understand and articulate what wellbeing is/should be and how this relates to SD.

Conclusion

Previous local indicator projects have often been set within a sustainable development paradigm and often had the twin aim of consultation and education. The focus has been on the local area: services, community relations, environment. This is well reflected in indicator sets like that produced by the Audit Commission (2002b: 2005) in the UK. However, my study shows that, when engaging policymakers and residents in an open discussion of wellbeing, rather than framing the discussion in terms of sustainable communities, then a wider and deeper view of wellbeing comes to the fore which, paradoxically, finds more synergy with sustainable development goals. Nurturing early emotional development, enjoying a more simple life, fostering relationships with the natural world, doing socially meaningful activities, building relationships of care, having a fairer society, valuing and supporting carers, less individualism and materialism and control over the private sector, were all regarded as important aspects of wellbeing but not reflected well in the AC set of local quality of life indicators. This is not surprising as many of these things are difficult to affect at the local level. Nor do they sit easily with dominant economic growth discourses as most of these goals, if attained by most people, may severely depress GDP. There we have a conundrum for local and central government and for society in general which the current economic system, based on promoting consumerism, is incapable of addressing. This is what Tim Jackson (2009) describes as 'the dilemma of growth'; more consumption causes environmental and social costs, less results in the risk of unemployment and recession. Considering the power and influence of global markets, making local government, communities and individuals responsible for wellbeing cannot solve this conundrum. Local policymakers in BVBC were caught by the same dilemma and, when given the opportunity to critically reflect, seemed unconvinced that the regeneration policies they were promoting would address long-term social justice and environmental issues. Undoubtedly aspects of life may be improved for many by addressing certain issues at a local or individual level. But if we want to promote *sustainable* wellbeing this means effectively tackling inequalities and addressing power relations which are propped up by current systems of politics and economy.

The current endeavour to measure what matters only goes so far. We must be less concerned about what is measurable and more concerned about what matters.

This poses enormous challenges on several levels. We must build institutional and social capacity to consider issues of power and legitimacy in debating different values. Many policymakers are personally reflective of their own values, as the case study shows, but this is not legitimised at an institutional level. This means increasing procedural justice, institutional capacity for reflexivity, dealing with conflict and respecting different values. Time, resources and understanding of the complexities and difficulties of this need to be built in to the process. The current focus on indicators as evidence produced by experts outside the policymaking process undermines the important discursive value of producing them. In addition, whilst policymakers are increasingly interested in qualitative research, such accounts struggle for legitimacy within the policy world where positivist paradigms of 'objective' quantitative evidence and data still hold sway. However, the importance of understanding everyday practices, norms and practical knowledge is important for developing effective policies. These things cannot always be translated into indicators.

A critical look at institutions and policy processes in this case study showed an arena which was very geared towards central discourses and responding to centrally set targets despite central government's desire to devolve responsibility. I concur with research which has already found that participation works in tightly controlled policy spaces which are responding to central agendas. Although these agendas are keen to mainstream ideas of participation, community, wellbeing and sustainability they were also in tension with neoliberal agendas which promote economic development and physical regeneration in order to attract affluent people and businesses to the area and are responding to dominant regional (feeding into national) economic indicators. Policy networks with decision making power were located around these agendas. These agendas provide the rational, 'realistic' and pragmatic basis against which differing values are judged.

However, discourses are dynamic and changing. This case study revealed the potential for indicator development to be a struggle between competing interests and throughout this struggle, capacity for deliberation was increased in terms of greater understanding of both wellbeing and sustainability. Recent national interest in subjective wellbeing influenced local discourses at Blyth Valley positively, producing creative controversies through which deeper discussions about the definition of both wellbeing and sustainability occurred. These particular accounts of wellbeing are also to some extent compatible with the tenets of sustainable development. A genuine focus on wellbeing, defined through democratic debate, contextual experience and supported by qualitative as well as quantitative research, has the potential to challenge neoliberal discourses of quality of life by promoting a less commodified, more socially-embedded view of development which could offer opportunities for sustainability discourses to find more legitimacy. However, we should be wary of a simplistic win-win discourse of subjective wellbeing and sustainability. While we live in a liberal society where individual freedom and justice is prized normatively and is one of the core foundations upon which our notions of wellbeing are built, this will continue to pose deep problems for a move to more ecologically sensitive forms of society. Although there are clear connections between subjective wellbeing and sustainability, there are also tensions. In addition, there is a danger that these discourses with their aim to support social capital and enhance community participation can be dove-tailed with central discourses of localism, active citizenship and community empowerment which are strongly critiqued in this book. These discourses have enormous power to 'co-opt' the subjective wellbeing agenda and some argue they are already doing so. So, while the 'new politics of wellbeing' is gaining momentum and has opened up spaces for discussion, as it did in this case study, we must also scrutinise which spaces it is closing down. A particular construct of wellbeing is being promoted as the real or true account of wellbeing, due to its scientific and objective basis. However this construct is created through a focus on individual subjective wellbeing and through a particular way of measuring this. This can only ever create a partial account. In a democratic society, we should be completely free to question this and argue for our own account based on our experience of life. Wellbeing is a complex human phenomenon and a political construct. As such, science will always struggle to represent what wellbeing is or what it should be. In contrast, each one of us can give both an expert opinion on what it is (as humans) and also on what it should be (as political citizens). We do not have to validate our experience or opinions against scientific accounts, although we can and should consider them as part of a healthy debate. What we should not do is allow them to automatically diminish our own expertise or opinions. Wellbeing measurement efforts may produce valuable knowledge and positive controversies and shifts, but we must also be vigilant about how they are privileged in discussions and what they undermine.

Policymakers therefore should take a genuine interest in a *democratically derived* account of wellbeing through working closely with the public and different interest and community groups, and appeals to evidence would be part of that debate. A stronger local democracy where different values are given respect and where conflicts are seen as productive democratic engagement must underpin policy. The development of indicators can act as an important deliberative process through which we can argue, listen, think, and learn in order to become clearer about what it is we value and how we want to measure it. This cannot be done by one person, or a team of researchers; nor must it rely only on 'consultation' or expert driven accounts.

What this book has aimed to deliver is the theoretical and evidential basis for arguing that wellbeing and sustainability are, above all, political projects, and what needs to underpin all sets of indicators is political debate about social values, where political participation *is* a social value, entrenched in the way we conceive of ourselves in the world and how we act on behalf of ourselves and others. This value of participation should run through all areas of life, in our work place, our home, our street, our meeting places virtual or real, as well as our democratic institutions. This is not a prescription to participate but it is a prescription to create a culture of participation where critical reflections on power relations are institutionalised and legitimised.

Appendix A

Audit Commission list of local quality of life indicators 2005

People and place

1   Priorities for improvement in the local area, as defined by local residents.

Community cohesion and involvement

2   The percentage of residents who think that people being attacked because of their skin colour, ethnic origin or religion is a very big or fairly big problem in their local area.

3   The percentage of residents who think that for their local area, over the past three years, community activities have got better or stayed the same.

4   Election turnout.

Community safety

5   The percentage of residents surveyed who said they feel 'fairly safe' or 'very safe' outside a) during the day; b) after dark.

6   **a)** Domestic burglaries per 1,000 households.
      b) Violent offences committed per 1,000 population.
      c) Theft of a vehicle per 1,000 population.
      d) Sexual offences per 1,000 population.

7   The percentage of residents who think that a) vandalism, graffiti and other deliberate damage to property or vehicles; b) people using or dealing drugs; and c) people being rowdy or drunk in public places is a very big or fairly big problem in their local area.

8   The number of a) pedestrian and b) cyclist road accident casualties per 100,000 population.

Culture and leisure

9   The percentage of the population within 20 minutes travel time (urban – walking, rural – by car) of different sports facility types.

10   The percentage of residents who think that for their local area, over the past three years the following have got better or stayed the same: a) activities for

teenagers; b) cultural facilities (for example, cinemas, museums); c) facilities for young children; d) sport and leisure facilities; and e) parks and open spaces.

Economic well-being

11 The percentage of the working-age population that is in employment.

12 **a)** The number of Job Seekers Allowance claimants as a percentage of the resident working-age population; and **b)** the percentage of these who have been out of work for more than a year.

13 **a)** The total number of VAT registered businesses in the area at the end of the year.
b) The percentage change in the number of VAT registered businesses.

14 Job density (number of jobs filled to working-age population).

15 The proportion of the population living in the most deprived super output areas in the country.

16 The percentage of the population of working age that is claiming key benefits.

17 The percentage of **a)** children and **b)** population over 60 that live in households that are income deprived.

Education and life-long learning

18 The percentage of half days missed due to total absence in **a)** primary and **b)** secondary schools maintained by the local education authority.

19 The proportion of young people (16- to 24-year-olds) in full-time education or employment.

20 The proportion of working-age population qualified to **a)** NVQ2 or equivalent; and **b)** NVQ4 or equivalent.

21 The percentage of 15-year-old pupils in schools maintained by the local authority achieving five or more GCSEs at grades A*–C or equivalent.

Environment

22 The proportion of developed land that is derelict.

23 The proportion of relevant land and highways that is assessed as having combined deposits of litter and detritus.

24 Levels of key air pollutants.

25 Carbon dioxide emissions by sector and per capita emissions.

26 Average annual domestic consumption of gas and electricity (kwh).

27 Daily domestic water use (per capita consumption).

28 The percentage of river length assessed as a) good biological quality; and b) good chemical quality.

29 The volume of household waste collected and the proportion recycled.

30 **a)** The percentage area of land designated as sites of special scientific interest (SSSI) within the local authority area in favourable condition; and
b) the area of land designated as a local nature reserve per 1,000 population.

Health and social well-being

31   Age-standardised mortality rates for **a)** all cancers; **b)** circulatory diseases; and **c)** respiratory diseases.

32   Infant mortality.

33   Life expectancy at birth (male and female).

34   The percentage of households with one or more person with a limiting long-term illness.

35   Teenage pregnancy, conceptions under 18 years, per 1,000 females aged 15–17.

Housing

36   The total number of new housing completions.

37   Affordable dwellings completed as a percentage of all new housing completions.

38   Household accommodation without central heating.

39   The percentage of residents who think that people sleeping rough on the streets or in other public places is a very big or fairly big problem in their local area.

40   The percentage of all housing that is unfit.

41   House price to income ratio.

Transport and access

42   The percentage of the resident population who travel to work **a)** by private motor vehicle; **b)** by public transport; and **c)** on foot or cycle.

43   The percentage of the resident population travelling over 20 km to work.

44   The percentage of residents who think that for their local area, over the past three years, **a)** public transport has got better or stayed the same; and **b)** the level of traffic congestion has got better or stayed the same.

45   Estimated traffic flows for all vehicle types (million vehicle km).

Other indicators

The indicators below cover important quality of life areas. Unfortunately, there are no guaranteed national data sources at present to provide comparable data for every local authority area. Nevertheless, we have listed them as we are confident that the indicators themselves are robust and that a national source is likely to become available in the next few years:

- The percentage of people surveyed who feel that their local area is a place where people from different backgrounds get on well together.
- The percentage of people surveyed who feel they can influence decisions affecting their local area.
- The percentage of people surveyed who find it easy to access key local services.
- The number of childcare places.

Appendix B

Martha Nussbaum's set of central human capabilities

1. Life

Being able to live to the end of a human life of normal length; not dying prematurely, or before one's life is so reduced as to be not worth living.

2. Bodily health

Being able to have good health, including reproductive health; to be adequately nourished; to have adequate shelter.

3. Bodily integrity

Being able to move freely from place to place; to be secure against violent assault, including sexual assault and domestic violence; having opportunities for sexual satisfaction and for choice in matters of reproduction.

4. Senses, imagination, and thought

Being able to use the senses, to imagine, think, and reason – and to do these things in a 'truly human' way, a way informed and cultivated by an adequate education, including, but by no means limited to, literacy and basic mathematical and scientific training.

Being able to use imagination and thought in connection with experiencing and producing works and events of one's own choice, religious, literary, musical, and so forth.

Being able to use one's mind in ways protected by guarantees of freedom of expression with respect to both political and artistic speech, and freedom of religious exercise. Being able to have pleasurable experiences and to avoid non-beneficial pain.

5. Emotions

Being able to have attachments to things and people outside ourselves; to love those who love and care for us, to grieve at their absence; in general, to love, to grieve, to experience longing, gratitude, and justified anger. Not having one's emotional development blighted by fear and anxiety. (Supporting this capability means supporting forms of human association that can be shown to be crucial in their development.)

6. Practical reason

Being able to form a conception of the good and to engage in critical reflection about the planning of one's life. (This entails protection for the liberty of conscience and religious observance.)

7. Affiliation

A. Being able to live with and toward others, to recognise and show concern for other human beings, to engage in various forms of social interaction; to be able to imagine the situation of another. (Protecting this capability means protecting institutions that constitute and nourish such forms of affiliation, and also protecting the freedom of assembly and political speech.)
B. Having the social bases of self-respect and non-humiliation; being able to be treated as a dignified being whose worth is equal to that of others. This entails provisions of non-discrimination on the basis of race, sex, sexual orientation, ethnicity, caste, religion, national origin.

8. Other species

Being able to live with concern for and in relation to animals, plants, and the world of nature.

9. Play

Being able to laugh, to play, to enjoy recreational activities.

10. Control over one's environment

A. Political. Being able to participate effectively in political choices that govern one's life; having the right of political participation, protections of free speech and association.
B. Material. Being able to hold property (both land and movable goods), and having property rights on an equal basis with others; having the right to seek employment on an equal basis with others; having the freedom from unwarranted search and seizure. In work, being able to work as a human being, exercising practical reason, and entering into meaningful relationships of mutual recognition with other workers.

Appendix C

Mapping exercise

Blyth Valley Wellbeing Framework	Martha Nussbaum Central Human Capabilities	Audit Commission Quality of Life Indicators
Personal qualities: Who we are		
Positive/thankful attitude to life	**4 Senses, imagination and thought.** Being able to…think and reason	
Philosophical approach/realistic expectations	**6 Practical reason.** Being able to form a conception of the good and to engage in critical reflection about the planning of one's life	
Sense of humour	**9 Play.** Being able to laugh…	
Inner peace/self-knowledge		
Being happy/content with life		
Emotional resilience/adaptability	**5 Emotions.** Not having one's emotional development blighted by overwhelming fear and anxiety	
Self-esteem/confidence	**7 Affiliation.** Having the social bases of self-respect and non-humiliation; being able to be treated as a dignified being whose worth is equal to that of others	
Initiative/motivation	**4 Senses, imagination and thought.** Being able to use imagination and thought in connection with experiencing and producing self-expressive works and events of one's own choice	
Being 'other-regarding': • Honest • Respectful • Caring • Understanding • Sharing • Generous	**7 Affiliation.** Being able to live with and toward others, to recognise and show concern for other human beings…to be able to imagine the situation of another and to have compassion for that situation; to have the capability for both justice and friendship	**Community cohesion** **2** The percentage of residents who think that people being attacked because of their skin colour, ethnic origin or religion is a very big or fairly big problem in their area **Community safety** **7** The percentage of residents who think that **a)** vandalism, graffiti and other deliberate damage to property or vehicles; **b)** people using or dealing drugs; and **c)** people being rowdy or drunk in public places is a very big or fairly big problem in their local area **Aspirational** The percentage of people surveyed who feel that their local area is a place where people from different backgrounds get on well together

Blyth Valley Wellbeing Framework	Martha Nussbaum Central Human Capabilities	Audit Commission Quality of Life Indicators
Health: How we are		**Health and social well-being**
General physical fitness and exercise	**1 Life.** Being able to live to the end of a human life of normal length; not dying prematurely	31 Age-standardised mortality rates for **a)** all cancers; **b)** circulatory diseases; and **c)** respiratory diseases
Healthy diet and healthy weight	**2 Bodily health.** Being able to have good health, including reproductive health; to be adequately nourished	32 Infant mortality
Reducing serious illness:		33 Life expectancy at birth (male and female)
• Cancer		34 The percentage of households with one or more person with a limiting long-term illness
• Cardio-vascular diseases		
• Respiratory diseases		
• Diabetes		
Help to alleviate the suffering caused by chronic pain/long-term illness:	**4 Senses, imagination and thought.** …to avoid non-necessary pain	34 The percentage of households with one or more person with a limiting long-term illness
• Accessibility of alternative therapies		
Drug, smoking, alcohol reduction		
Mental health	**5 Emotions.** Not having one's emotional development blighted by overwhelming fear or anxiety, or by traumatic events of abuse or neglect	**Culture and leisure**
Opportunity to rest, relieve stress and recover in natural/tranquil surroundings	**4 Senses, imagination and thought.** …being able to have pleasurable experiences	10 The percentage of people who think that for their local area, over the past three years the following have got better or stayed the same
	8 Other species. Being able to live with concern for and in relation to animals, plants and the world of nature.	… **e)** parks and open spaces
		Environment
		30 **a)** The percentage area of land designated as sites of special scientific interest (SSSI) within the local authority area in favourable condition; and **b)** the area of land designated as a local nature reserve per 1,000 population

Blyth Valley Wellbeing Framework	Martha Nussbaum Central Human Capabilities	Audit Commission Quality of Life Indicators
Support for elderly: • Dignity and respect for individuals • Quality and affordability of care for elderly • Support for people to be cared for at home or with family Support for carers Quality of GP and health services Quality of death and support for those bereaved	**7 Affiliation.** Having the social bases of self-respect and non-humiliation; being able to be treated as a dignified being whose worth is equal to that of others	
Activity: What we do Keeping active and busy: • Feeling challenged/stimulated • Being absorbed in something • Something to look forward to Ability to pursue interests/hobbies/play Provision of activities/facilities Being able to enjoy 'simple things'	**4 Senses, imagination and thought.** Being able to use imagination and thought in connection with experiencing and producing self-expressive works and events of one's own choice, religious, literary, musical and so forth **9 Play.** Being able to laugh, to play, to enjoy recreational activities	**Culture and leisure** **10** The percentage of people who think that for their local area, over the past three years, the following have got better or stayed the same: **a)** activities for teenagers; **b)** cultural facilities (for example, cinemas, museums); **c)** facilities for young children; **d)** sport and leisure facilities; and **e)** parks and open spaces
Availability of jobs Quality of job: • Job satisfaction • Being valued • Good working relationships • Less commuting • Flexibility of hours/part-time work/home-working • Opportunities to train/progress • Opportunities and support for self-employment	**7 Affiliation.** In work, being able to work as a human being, exercising practical reason and entering into meaningful relationships of mutual recognition with other workers **10 Control over one's environment. B. Material.** Having the right to seek employment on an equal basis with others	**Economic wellbeing** **11** The percentage of the working-age population that is in employment **13 a)** The total number of VAT registered businesses in the area at the end of the year. **b)** The percentage change in the number of VAT registered businesses **14** Job density (number of jobs filled to working-age population)

Blyth Valley *Wellbeing Framework*	Martha Nussbaum *Central Human Capabilities*	Audit Commission *Quality of Life Indicators*
Quality of education, including: • Life skills/social values • Numeracy and literacy for all • Supporting child's interests • Accommodating child's needs • Valuing practical skills • Academic achievement • Access to lifelong learning • Opportunities to (re)train	**4 Senses, imagination and thought.** …Adequate education, including, but by no means limited to, literacy and basic mathematical and scientific training	**Education and lifelong learning** **18** The percentage of half days missed due to total absence in **a)** primary; and **b)** secondary schools maintained by the local education authority **19** The proportion of young people (16- to 24-year-olds) in full-time education or employment **20** The proportion of the working-age population qualified to **a)** NVQ2 or equivalent; and **b)** NVQ4 or equivalent **21** The percentage of 15-year-old pupils in schools maintained by the local authority achieving five or more GCSEs at grades A*–C, or equivalent
Income: How we manage financially Adequate income for: • Maintaining a comfortable home • Meeting family/social needs • Healthy diet and lifestyle • Meeting health/care needs • Leisure and cultural pursuits • Annual holiday Economic security: • Ability to plan for future life • Ability to provide for family • Ability to insure against misfortune • Ability to save Control over debt Fair distribution of wealth		**Economic wellbeing** **12 a)** The number of Job Seekers Allowance claimants as a percentage of the resident working-age population; and **b)** the percentage of these who have been out of work for more than a year **15** The proportion of the population living in the most deprived areas in the country **16** The percentage of the population of working age claiming key benefits **17** The percentage of **a)** children and **b)** the population over 60 that live in households that are income deprived

Blyth Valley Wellbeing Framework	Martha Nussbaum Central Human Capabilities	Audit Commission Quality of Life Indicators
Social world: How we relate to each other		
Loving relationships with family/friends: • Stability of relationships • Giving and receiving support and encouragement • Opportunities to spend happy times together without stress	**5 Emotions.** Being able to have attachments to things and people outside ourselves; to love those who love and care for us… (Supporting this capability means supporting forms of human association that can be shown to be crucial in their development)	**Community cohesion and involvement** **3** The percentage of residents who think that for their local area, over the past three years, community activities have got better or stayed the same
Ability to care for others/opportunities for voluntary work		**Aspirational** The percentage of people surveyed who feel that their local area is a place where people from different backgrounds get on well together
Community spirit/compromise: • Neighbours helping each other • Quality/stability of relationships • Sense of belonging • Knowing lots of people in area • Balance between privacy and support • Ability to tolerate differences	**7 Affiliation.** Being able to live with and toward others, to recognise and show concern for other human beings, to engage in various forms of social interaction… (Protecting this capability means protecting institutions that constitute and nourish such forms of affiliation)	
General friendliness of area: • Levels of trust • Hospitality/helpfulness to strangers/visitors • Cultural diversity celebrated • Sense of identity and pride		
Collective parenting/clear boundary setting for young people		
Good intergenerational relationships/mutual respect		
Mutual interest associations		**Community cohesion and involvement** **3** The percentage of residents who think that for their local area, over the past three years, community activities have got better or stayed the same

Blyth Valley Wellbeing Framework	Martha Nussbaum Central Human Capabilities	Audit Commission Quality of Life Indicators
Physical world: How we relate to our surroundings		
Clean water, air and land		**Environment** 22 The proportion of developed land that is derelict 27 Daily domestic water use (per capita consumption) 28 The percentage of river length assessed as **a)** good biological quality; and **b)** good chemical quality 24 Levels of key air pollutants
Tranquillity/peace and quiet Beauty and diversity (of built and natural world) Opportunities to experience the natural world and other species, landscapes, flora and fauna, birdsong, pets	**8 Other species.** Being able to live with concern for and in relation to animals, plants, and the world of nature	**30 a)** The percentage area of land designated as sites of special scientific interest (SSSI) within the local authority area in favourable condition; and **b)** the area of land designated as a local nature reserve per 1,000 population
Comfortable and affordable homes for all (to rent or buy) Housing security	**2 Bodily health.** To have adequate shelter **10 Control over one's environment. B. Material.** Being able to hold property (both land and moveable goods), not just formally but in terms of real opportunity…and having property rights on an equal basis with others	**Housing** 36 The total number of new housing completions 37 Affordable dwellings completed as a percentage of all new housing completions 38 Household accommodation without central heating 39 The percentage of residents who think that people sleeping rough on the streets or in other public places is a very big or fairly big problem in their local area 40 The percentage of all housing that is unfit 41 House price to income ratio

Blyth Valley *Wellbeing Framework*	Martha Nussbaum *Central Human Capabilities*	Audit Commission *Quality of Life Indicators*
Clean, peaceful, safe neighbourhood		**Community cohesion** **2** The percentage of residents who think that people being attacked because of their skin colour, ethnic origin or religion is a very big or fairly big problem in their local area **Community safety** **5** The percentage of residents surveyed who said they feel 'fairly safe' or 'very safe' outside **a)** during the day; and **b)** after dark **6 a)** Domestic burglaries per 1,000 households **b)** Violent offences committed per 1,000 population **c)** Theft of a vehicle per 1,000 population **d)** Sexual offences per 1,000 population **7** The percentage of residents who think that **a)** vandalism, graffiti and other deliberate damage to property or vehicles; **b)** people using or dealing drugs; and **c)** people being rowdy or drunk in public places, is a very big or fairly big problem in their local area **Environment** **23** The proportion of relevant land and highways that is assessed as having combined deposits of litter and detritus
Access to facilities and services: Good GP service • Parks • Play facilities • Libraries • Sports centres		**Culture and leisure** **9** The percentage of the population within 20 minutes travel time (urban – walking, rural – by car) of different sports facility types **Aspiration** **3** The percentage of people surveyed finding it easy to access key local services
Waste disposal and recycling		**Environment** **29** The volume of household waste collected and the proportion recycled

Blyth Valley Wellbeing Framework	Martha Nussbaum Central Human Capabilities	Audit Commission Quality of Life Indicators
Transport and mobility: • Road safety • Reduction in cars • Traffic control		**Transport and access** 42 The percentage of the resident population who travel to work **a)** by private motor vehicle; **b)** by public transport; and **c)** on foot or cycle 43 The percentage of the resident population travelling over 20 km to work 44 The percentage of residents who think that for their local area, over the past three years, **a)** public transport has got better or stayed the same; **b)** the level of traffic congestion has got better or stayed the same 45 Estimated traffic flows for all vehicle types (million vehicle km) **People and place** 8 The number of **a)** pedestrian and **b)** cyclist road accident casualties per 100,000 population
Renewable energy/reduction in fuel poverty Reducing climate change		**Environment** 25 Carbon dioxide emissions by sector and per capita emissions 26 Average annual domestic consumption of gas and electricity (kwh)
Freedom: Being in control of one's own life Being able to live the life you choose	4 **Senses, imagination, and thought.** Being able to search for the ultimate meaning of life in one's own way	
Freedom from stereotyping, unfair discrimination, intimidation, abuse and violence, inside and outside the home: • Respect for life choices • Protection from abuse • Provision of places for gathering • Recognition of festivals/rituals • Social norms of speaking out and standing up against abuse or injustice	3 **Bodily integrity.** Being able to move freely from place to place; having one's bodily boundaries treated as sovereign, i.e. being able to be secure against assault, including sexual assault, child sexual abuse, and domestic violence 7 **Affiliation. B.** Having the social bases of self-respect and non-humiliation; being able to be treated as a dignified being whose worth is equal to that of others. This entails, at a minimum, protections against discrimination on the basis of race, sex, sexual orientation, religion, caste, ethnicity, or national origin	**Community cohesion** 2 The percentage of residents who think that people being attacked because of their skin colour, ethnic origin or religion is a very big or fairly big problem in their local area **Community safety** 5 The percentage of residents surveyed who said they feel 'fairly safe' or 'very safe' outside **a)** during the day; and **b)** after dark 6 …**b)** Violent offences committed per 1,000 population…**d)** Sexual offences per 1,000 population **Aspirational** The percentage of people surveyed who feel that their local area is a place where people from different backgrounds get on well together

Blyth Valley *Wellbeing Framework*	Martha Nussbaum *Central Human Capabilities*	Audit Commission *Quality of Life Indicators*
Fairness in how people are treated Equality of opportunity: • Having life chances/confidence • Access to resources and support to live the life one chooses • Accommodating different needs • Access to childcare • Support for people with disability to access work	**7 Affiliation. B.** Having the social bases of self-respect and non-humiliation; being able to be treated as a dignified being whose worth is equal to that of others. This entails, at a minimum, protections against discrimination on the basis of race, sex, sexual orientation, religion, caste, ethnicity, or national origin **10 Control over one's environment. B. Material.** Having property rights on an equal basis with others; having the right to seek employment on an equal basis with others	
Participation in and influence on local decision-making: • Freedom to/forum for debate • Openness of decision-making • Procedural fairness • Opportunity/support for collective action • Local control over budgets. Women and minority groups represented Promotion of and access to independent support, advice and advocacy for addressing problems, injustices and abuse	**7 Affiliation.** …Protecting the freedom of assembly and political speech **10 Control over one's environment. A. Political.** Being able to participate effectively in political choices that govern one's life; having the right of political participation, protections of free speech and association	**Community cohesion and involvement** **4** Election turnout **Aspirational** The percentage of people surveyed who feel they can influence decisions affecting their local area **Aspirational** The percentage of people surveyed finding it easy to access key local services

Notes

1 Introduction

1 Millennium Ecosystem Assessment, 2005. *Ecosystems and Human Well-being: Synthesis*. Island Press, Washington, DC.

2 Marmot, M. (2011) *Fair Society, Healthy Lives*. Report to The Department of Health http://www.marmotreview.org/ Wilkinson, R. and Pickett, K. (2010) *The Spirit Level* Penguin Books. London.

3 The so-called Easterlin Paradox of economic growth; statistics across developed nations show that whilst GDP has doubled in the post-war period, life satisfaction has flatlined. For a review see Offer, A. (2000) *Economic Welfare Measurements and Human Wellbeing*. Discussion Papers in Economic and Social History No 34. University of Oxford.

4 See O'Neill (1993) Chapter 6 for a review of Aristotle's philosophies.

5 Michalos *et al.* (2011: 2).

6 For the moment these two terms are used interchangeably although discussion of separate definitions, meanings and understandings of each term follows later in this chapter and in Chapter 2.

2 Human wellbeing and quality of life

1 For a historical discussion of the pursuit of the good life see Schoch (2006).

2 The difference between GNP and GDP is that GNP counts the economic production of all nationals whether they are in the country or not. GDP focuses on economic activity within a country's borders. The USA switched from GNP to GDP in 1991 and many other countries have done so.

3 In utilitarian literature and in economics 'welfare' is synonymous with happiness, life satisfaction and 'utility'. However, some authors, like Jordan (2008) seek to distinguish welfare from wellbeing, arguing that the former is too narrowly defined by the distribution of material resources. The latter he equates to a notion of 'social value' which is distinct from social capital (which he argues adds nothing to current economic models of welfare and even uses market terminology).

4 As philosopher Martha Nussbaum says, 'there are many species of utilitarianism' (2005: 32) and the explanation presented here is necessarily brief. Essentially, the theory holds that if a greater number of people can satisfy their individual preferences, the overall level of utility (happiness) in society will increase.

5 Of course, social statistics and indicators were not new. The rise of statisticians in the 18th century both responded to and helped define the concept of 'population' and its corresponding measurement and management (Foucault 1991a). See Cobb and Rixford (2005) for an account of indicators through the 19th and 20th centuries in the

USA leading up to the social indicator movement. Nevertheless, the 1960s and 1970s were distinct because of the prominence and proliferation of social reporting that went on in several countries.

6 "I think we've been through a period where too many people have been given to understand that if they have a problem, it's the government's job to cope with it. 'I have a problem, I'll get a grant.' 'I'm homeless, the government must house me.' They're casting their problem on society. And, you know, there is no such thing as society. There are individual men and women, and there are families. And no government can do anything except through people, and people must look to themselves first. It's our duty to look after ourselves and then also to look after our neighbour. People have got the entitlements too much in mind, without the obligations. There's no such thing as entitlement, unless someone has first met an obligation." Margaret Thatcher, talking to *Women's Own* magazine, October 31, 1987.

7 Rawls identified two different categories of 'primary goods': 'basic rights and liberties', which should be accorded to each individual equally; and other 'primary goods', which could be distributed unequally if necessary but with provisos (Rawls, 1971).

8 Although Rawls' idea of 'goods' was wide, including rights and liberties.

9 For example, Bristol City Council started in 2006 to include a whole host of questions around love, support, opportunities.

10 There is not room here to go into detail but many psychologists make a distinction between happiness and positive affect/emotion research, the latter focusing on more specific states including excitement, pleasure, alertness, joy (see Forgeard *et al,* 2011; Seligman 2011).

11 For example by increasingly influential organisations like the New Economics Foundation.

12 This book is one of many in the popular press which try to promote the old idea that 'money doesn't buy happiness'. One of the ways it does this is to describe the lives of various miserable rich people around the world. Whilst SWB research shows a complex relationship between money and SWB, it does not show that money makes individuals miserable, just that it has rapidly diminishing returns at a statistical level. The relationship is complex depending on each person's values, what they do with their money and their relative position in society.

3 Sustainable wellbeing: an oxymoron?

1 According to current Intergovernmental Panel on Climate Change targets, see http://www.ipcc.ch/publications_and_data/ar4/syr/en/contents.html.

2 For example European Commission, OECD and other partners 'Beyond GDP Initiative' http://www.beyond-gdp.eu/ and Stiglitz, J.E, Sen, A. and Fitoussi, J.P. (2009) Report by the *Commission on the Measurement of Economic Performance and Social Progress* OECD. http://www.stigliz-sen-fitoussi.fr/.

3 For example see Stern, N. (2006*) The Economics of Climate Change: The Stern Review.* Cambridge University Press.

4 The then Norway Minister Gro Harlem Brundtland was chair of the commission which created the landmark definition of sustainable development: 'Development that meets the needs of the present without compromising the ability of future generations to meet their own needs' (WCED, 1987).

5 For example, Dryzek, 1997; Dresner, 2002; Connolly, 2007.

6 Although much of the literature focuses on the advanced liberal democracies of the USA, Canada, Europe and Australia.

7 With the New Economics Foundation and the University of West England.

4 Leave it to the people?

1 Baker, E. (transl.) (1946) *The Politics of Aristotle*. Clarendon Press: Oxford.
2 Miller *et al.* (2000) found that local people's perception of a 'citizen' tended to include everyone who lived in an area 'no matter how recently arrived or how shallow their roots' (p.4). This is how I also use the term for the purposes of this discussion.
3 The Rio Declaration on Environment and Development, United Nations Conference on Environment and Development, Rio de Janeiro, 1992.
4 The Guardian October 2, 2001.
5 Although this began to happen with the new Local Area Agreements, national targets partly dictated how the money should be spent.

5 The role of indicators

1 Although, as already argued, this is heavily guided by the construct of 'sustainable communities'.

6 Case study of Blyth Valley Borough Council

1 *The Reshaping of British Railways* (1963) by Richard Beeching was a report to UK government on the modernisation of the railway system and resulted in the closure of 50% of branch stations over the next decade.
2 SENNTRi ceased to exist in 2009 following local government restructuring and was replaced by the South East Northumberland Regeneration team, which is fully funded by Northumberland County Council and will carry out the regeneration roles previously delivered by former district authorities and SENNTRi.
3 The UK government's national regeneration agency.
4 Gross value added (GVA) is a measure of economic activity which contributes to the estimation of GDP.
5 Information taken from BVBC's own promotional literature.
6 Determined by the Index of Multiple Deprivation (IMD).
7 Except for the Isles of Scilly, Greater London and six metropolitan counties.
8 This perception is not entirely well reflected in election results which show variable support for the Conservative Party over the past two decades across the region (Randall 2009).
9 Liberal Democrats 26; Conservatives 19; Labour 17; Independent 5.
10 A national biodiversity designation.
11 Women's Issues Group. Minutes of meeting held 9 February 2006, Stannard Room, Blyth Civic Centre.
12 When I first arrived the cabinet had one woman councillor in it but she passed away a year later.

7 Defining 'local' wellbeing

1 Taken from an email written by the sustainability officer to colleagues asking for input into these indicators, 13/2/04.
2 Quality of Life Indicators Progress Report to Blyth Valley LSP Board, October 2005.
3 Blyth Market; Blyth Sports Centre; Blyth beach; Cramlington Manor Walks indoor shopping centre; Cramlington Village; Concordia Leisure Centre; and Seaton Delaval Yourlink (council one stop shop).
4 The sustainability officer helped with the production of this report through many conversations, feedback comments on early drafts and helping to structure the final draft.

8 Developing a wellbeing framework

1 'Charvers' is a derogatory term for a stereotyped group of young people identified through their clothes, taste in music, social class and behaviour.
2 This may have been because the focus of the discussion was on their own wellbeing and it should not be assumed that only a few members of the public cared about these issues.

References and further reading

Abdallah, S., Mahony, S., Marks, N., Michaelson, J., Seaford, C., Stoll, L. and Thompson, S. (2011) *Measuring Our Progress: The Power of Wellbeing*. London: New Economics Foundation.

Adamson, D. and Bromiley, R. (2008) *Community Empowerment in Practice: Lessons from Communities First*. York: Joseph Rowntree Foundation.

Agrawal, A. (2005) *Environmentality: Technologies of Government and Political Subjects*. Durham, NC: Duke University Press.

Alkire, S. (2002) Dimensions of Human Development, *World Development*, 30 (2): 181–205.

Alkire, S. (2005) *Capability and Functionings: Definition and Justification*. Human Development and Capability Association Briefing http://www.capabilityapproach.com/pubs/HDCA_Briefing_Concepts.pdf (accessed 12/01/08).

Amin, A. (2005) Local community on trial, *Economy and Society*, 34(4): 612–633.

Anand, P., Hunter, G. and Smith, R. (2005) Capabilities and Well-Being: Evidence Based on the Sen–Nussbaum Approach to Welfare, *Social Indicators Research*, 74(1), 9–55.

Anand, P., Santos, C. and Smith, R. (2007a) The Measurement of Capabilities. Open Discussion papers in Economics 67, Open University UK http://www.open.ac.uk/socialsciences/__assets/cg47myi2epp7enzgdp.pdf (accessed 18/2/2008).

Anand, P., Hunter, G., Carter, I., Dowding, K., Guala, F. and van Hees, M. (2007b) Measurement of Human Capabilities. Open Discussion Papers in Economics 53, Open University UK http://www.open.ac.uk/socialsciences/__assets/5ih6kr7z8lqdcz8ayd.pdf (accessed 18/2/2008).

Anand, P., Hunter, G., Carter, I., Dowding, K., Guala, F. and Van Hees, M. (2009) The Development of Capability Indicators, *Journal of Human Development and Capabilities*, 10(1): 125–152.

Arendt, H. (1958) *The Human Condition*. Chicago: University of Chicago Press.

Arnstein, S. R. A. (1969) Ladder of Citizen Participation, *Journal of the American Planning Association*, 35(4): 216–224.

Astleithner, F. and Hamedinger, A. (2003) The analysis of sustainability indicators as socially constructed policy instruments: benefits and challenges of interactive research, *Local Environment*, 8(6): 627–640.

Astleithner, F., Hamedinger, A., Homan, N. and Rydin, Y. (2004) Institutions and indicators – The discourse about indicators in the context of sustainability, *Journal of Housing and the Built Environment*, 19: 7–24.

Atkinson, S. and Joyce, K. E. (2011) The place and practices of well-being in local governance, *Environment and Planning C: Government and Policy*, 29(1): 133–148.

Audit Commission (2002a) *Audit Commission Quality of Life Indicators Pilot 2001/2002.* HMSO.

Audit Commission (2002b) *Quality of Life: Using Quality of Life Indicators.* HMSO.

Audit Commission, Local Government Association, MORI, New Economics Foundation (2002c unpublished) *Quality of Life Pilots Survey. Final Topline Results, 23 May.*

Audit Commission (2003) *Quality of Life: A good practice guide to communicating quality of life indicators.* Audit Commission, London. http://www.audit-commission.gov.uk/ Products/NATIONAL-REPORT/ACDE5F73-1CEB-4936-9675-6FD06664CC7F/ QualityofLifeGoodpracticeguide.pdf (accessed 8/11/04).

Audit Commission (2004) *Comprehensive Performance Assessment Blyth Valley Borough Council.* Audit Commission http://www.audit-commission.gov.uk/Products/ CPA-DISTRICT-REPORT/49DAC178-A841-496c-9BB4-F7361555BDA3/ BlythValleyCPA8Apri04.pdf (accessed 11/2/2008).

Audit Commission (2005) *Local quality of life indicators – supporting local communities to become sustainable.* London, Audit Commission.

Ayong Le Kama, A. (2007) Preface, *International Journal of Sustainable Development (Special Edition on Sustainability Indicators)*, 10(1–2), 4–13.

Bache, I. and Reardon, L. (2011) An idea whose time has come? Explaining the rise of well-being in British politics, paper presented at the Department of Politics, University of Sheffield, November 17, 2011.

Baker, S. (2006) *Sustainable Development.* Abingdon: Routledge.

Baldwin, A. (2003) The nature of boreal forest, *Space and Culture*, 6: 415–428.

Ballas, D., Dorling, D. and Shaw, M. (2007) Societal Inequality, Health and Well-Being, in Howarth, J. and Hart, G. (eds) *Well-Being: Individual, Community and Social Perspectives*, Basingstoke: Palgrave Macmillan.

Barnes, M., Newman, J. and Sullivan, H. (2004) Power, Participation and Renewal: Theoretical Perspectives on Public Participation under New Labour in Britain, *Social Politics*, 11(2), 267–279.

Barnett, N. (2011) Local Government at the Nexus? *Local Government Studies*, 37(3): 275–290.

Bauman, Z. (2008) *The absence of society*, Joseph Rowntree Foundation Viewpoint Social Evils Series, York: Joseph Rowntree Foundation.

Bayliss, D. and Walker, G. (1996) Environmental Monitoring and Planning for Sustainability, in Buckingham-Hatfield, S. and Evans, B. *Environmental Planning and Sustainability*, Chichester: Wiley.

BBC News (1982) *UK unemployment tops three million*, On This Day, BBC website. http:// news.bbc.co.uk/onthisday/hi/dates/stories/january/26/newsid_2506000/2506335.stm (accessed 14/2/08).

Beatty, C. and Fothergill, S. (2005) The Diversion from 'Unemployment' to 'Sickness' across British Regions and Districts. *Regional Studies*, 39(7): 837–854.

Beatty, C., Fothergill, S. and Powell, R. (2007) Twenty years on: has the economy of the UK coalfields recovered. *Environment and Planning A*, 39: 1654–1675.

Beck, U. (1992) *Risk Society: Towards a New Modernity.* London: Sage.

Beck, W., van der Maesen, L. and Walker, A. (eds) (1997) *The Social Quality of Europe*, The Hague, Kluwer Law International.

Beck, W., van der Maesen, L., Thomése, F. and Walker, A. (eds) (2001a) *Social Quality: A Vision for Europe*, The Hague, Kluwer Law International.

Beck, W., van der Maesen, L. and Walker, A. (2001b) Theorizing social quality: the concept's validity, in Beck, W., van der Maesen, L., Thomése, F. and Walker, A. (eds) *Social Quality: A Vision for Europe*, The Hague, Kluwer Law International.

Bell, D. (2005) Annex 1: Review of Research into Subjective Well-being and its Relation to Sport and Culture, in *Quality of Life and Well-being: Measuring the Benefits of Culture and Sport: Literature Review and Thinkpiece*. Glasgow: Scottish Executive Social Research

Bell, S. and Morse, S. (2000) *Sustainability Indicators: Measuring the Immeasurable*. London: Earthscan.

Bell, S. and Morse, S. (2003) *Measuring Sustainability: Learning by Doing*. London: Earthscan.

Bennett, K., Benyon, H. and Hudson, R. (2000) *Coalfields Regeneration: Dealing with the Consequences of Industrial Decline*. Bristol: Policy Press.

Blowers, A. (2003) Inequality and Community: The Missing Dimensions of Sustainable Development, in Buckingham, S. and Theobald, K. *Local Environmental Sustainability*. Cambridge: Woodhead Publishing.

Blumer, H. (1971) Social Problems as Collective Behaviour. *Social Problems*, 18: 298–306.

Boaden, N., Goldsmith, M., Hampton, W. and Stringer, P. (1982) *Public Participation in Local Services*. Harlow: Longman.

Boulanger, P. (2007) Political uses of social indicators: overview and application to sustainable development indicators. *International Journal of Sustainable Development*, 10(1–2), 14–32.

Boyne, G. A. (1999) Processes, Performance and Best Value in Local Government. *Local Government Studies*, 25(2): 1–15.

Browne, A. (2008) *Has there been a decline in values in British society?* The Social Evils Series Viewpoint Paper. York: Joseph Rowntree Foundation. http://www.jrf.org.uk/publications/has-there-been-decline-values-british-society (accessed August 2011).

Buckingham-Hatfield, S. and Evans, B. (1996b) *Environmental Planning and Sustainability*. Chichester: Wiley.

Buckingham-Hatfield, S. and Evans, B. (1996a) Achieving Sustainability through Environmental Planning, in Buckingham-Hatfield, S. and Evans, B. *Environmental Planning and Sustainability*. Chichester: Wiley.

Burningham, K. and Thrush, D. (2001) '*Rainforests are a long way from here': The environmental concerns of disadvantaged groups*. York: Joseph Rowntree Foundation.

Bush, J. (2005) *Quality of Life and Participation in Neighbourhood Renewal in Ex-Mining Communities in County Durham*. ESRC/ODPM Post-Doctoral Fellowship Research Report, School of Population and Health Sciences, Newcastle University.

BVBC, Blyth Valley Borough Council (2007) *Blyth Valley Strategic Objectives 2008-2025*. Blyth, Newcastle.

Cahill, C. (2007) Well positioned? Locating participation in theory and practice, *Environment and Planning A*, 39: 2861–2865.

Cameron, D. (2010) Speech to House of Commons on measuring wellbeing, 25 November 2010. http://www.number10.gov.uk/news/pm-speech-on-well-being/ (accessed 15 September 2011).

Canoy, M., and Lerais, F. (2007) Beyond GDP: Overview paper for the Beyond GDP conference. Bureau of European Policy Advisors (BEPA): European Commission. http://www.beyond-gdp.eu (accessed December 2008).

Carlisle, S. and Hanlon, P. (2007) Well-being and consumer culture: a different kind of public health problem? *Health Promotion International*, 22(3): 261–268.

Carman, C. (2010) The Process is the Reality: Perceptions of Procedural Fairness and Participatory Democracy, *Political Studies*, 58: 731–751.

Carmichael, J., Talwar, S., Tansey, J. and Robinson, J. (2005) Where Do We Want To Be? Making Sustainability Indicators Integrated, Dynamic and Participatory, in Phillips, R. (ed.) *Community Indicators Measuring Systems*. Aldershot and Burlington: Ashgate.

Chambers, N., Simmons, C. and Wackernagel, M. (2000) *Sharing Nature's Interest: ecological footprints as an indicator of sustainability.* London: Earthscan.

Cheshire, P. (2007) *Segregated Neighbourhoods and Mixed Communities: a critical analysis.* York: Joseph Rowntree Foundation.

Clary, J., Dolfsma, W. and Figart, D. (2006) *Ethics and the Market: Insights from Social Economics.* Abingdon: Routledge.

Cobb, C. (2000) *Measurement Tools and the Quality of Life*, Redefining Progress. http://www.rprogress.org/publications/2000/measure_qol.pdf (accessed 18/2/2008).

Cobb, C. and Rixford, C. (2005) Historical Background of Community Indicators, in Phillips, R. (ed.) *Community Indicators Measuring Systems*, Aldershot and Burlington: Ashgate.

Cohen, G. A. (1993) Equality of What? On Welfare, Goods and Capabilities, in Nussbaum, M. and Sen, A. (eds) *The Quality of Life.* New York: Oxford University Press.

Cohen, M. D., March, J. G. and Olsen, J.P. (1972) A Garbage Can Model of Organizational Choice, *Administrative Science Quarterly*, 17(1): 1–25.

Collingridge, D. and Reeve, C. (1986) *Science Speaks to Power: The Role of Experts in Policy Making.* London: Frances Pinter.

Connelly, S. (2007) Mapping Sustainable Development as a Contested Concept, *Local Environment*, 12(3): 259–278.

Cook, T. E., and Morgan, P. M. (1971) *Participatory Democracy*, San Francisco: Canfield Press.

Cooke, B. (2001) The Social Psychological Limits of Participation, in Cooke, B. and Kothari, U. *Participation: The New Tyranny?* London, New York: Zed Books.

Cooke, B. and Kothari, U. (2001) *Participation: The New Tyranny?* London, New York: Zed Books.

Cornwall, A. (2004) New democratic spaces? The politics and dynamics of institutionalised participation, *IDS Bulletin*, 35(2): 1–10.

Cronin de Chavez, A., Blackett-Milburn, K., Parry, O. and Platt, S. (2005) Understanding and researching wellbeing: Its usage in different disciplines and potential for health research and health promotion, *Health Education Journal*, 64(1): 70–87.

Dalton, R. J. (2004) *Democratic Challenges, Democratic Choices.* Oxford: Oxford University Press.

Darlow, A., Percy-Smith, J. and Wells, P. (2007) Community Strategies: Are They Delivering Joined Up Governance? *Local Government Studies*, 33(1): 117–129.

Dasgupta, P. (2004) *Human Well-Being and the Natural Environment.* Oxford: Oxford University Press.

Davidson, S. (1998) Spinning the Wheel of Empowerment. *Planning*, 1262, April 3.

Defra (2005) *Securing the Future: delivering UK sustainable development strategy.* Norwich: The Stationary Office.

Defra (2006) *Sustainable Communities: A Shared Agenda, A Share of the Action. A Guide for Local Authorities.* Defra, London. http://www.sustainable-development.gov.uk/publications/documents/sustainable-communities-guide.pdf (accessed 3/4/2007).

Defra (2007) *Common Understanding of Wellbeing for Policy.* London: Defra http://www.defra.gov.uk/sustainable/government/what/priority/wellbeing/common-understanding.htm.

Department of Communities and Local Government website *What is a Sustainable Community?* webpage. http://www.communities.gov.uk/communities/sustainable communities/whatis/.

Dery, D. (2000) Agenda Setting and Problem Definition, *Policy Studies*, 21(1): 37–47.

DETR (1998) *Modernising Local Government: In Touch with the People* White Paper, DETR London.

DETR (1999) *A Better Quality of Life: A strategy for sustainable development for the United Kingdom.* London: The Stationary Office.

DETR (2000) *Local Quality of Life Counts: A handbook for menu of local indicators of sustainable development.* HMSO.

Diener, E. (2000) Subjective Well-Being: The science of happiness, and a proposal for a national index, *American Psychologist*, 55: 34–43.

Diener, E. and Oishi, S. (2000) Money and happiness: Income and subjective well-being across nations, in E. Diener and E. M. Suh (eds), *Subjective Well-being across Cultures.* Cambridge, MA: MIT Press.

Diener, E. and Seligman, M. (2004) Beyond money: toward an economy of well-being, *Psychological Science in the Public Interest*, 5: 1–31.

Diener, E., Lucas, R., Schimmack, U. and Helliwell, J. (2009) *Well-being for Public Policy*, New York: Oxford University Press.

Dinham, A. (2005) Empowered or over-powered? The real experiences of local participation in the UK's New Deal for Communities, *Community Development Journal*, 40(3), 301–312.

Dinham, A. (2006) Raising expectations or dashing hopes?: Well-being and participation in disadvantaged areas, *Community Development Journal*, 42(2): 181–193.

Dolan, P., Peasgood, T. and White, M. (2006b) *Review of research on personal well-being and application to policy making.* London: DEFRA.

Dolan, P., Peasgood, T., Dixon, A., Knight, M., Phillips, D., Tsuchiya, A. and White, M. (2006a) *Research on the relationship between well-being and sustainable development.* London: DEFRA.

Dolan, P., Peasgood, T. and White, M. (2008) Do we really know what makes us happy? A review of the economic literature on the factors associated with subjective well-being, *Journal of Economic Psychology* 29, 94–122.

Donovan, N. and Halpern, D. (2002) *Life Satisfaction: the state of knowledge and the implications for government.* London: Prime Minister's Strategy Unit.

Dorn *et al.* (2007) Is it Culture or Democracy? The Impact of Democracy and Culture on Happiness, *Social Indicators Research* 82, 505–526.

Doyal, L. and Gough, I. (1991) *A Theory of Human Need*, Basingstoke: MacMillan.

Dresner, S. (2002) *The Principles of Sustainable Development.* London: Earthscan.

Dryzek, J. S. (1997) *The Politics of the Earth: Environmental Discourses.* New York: Oxford University Press.

Dworkin, R. (1981b) What is Equality? Part 1: Equality of Welfare, *Philosophy and Public Affairs* 10, 185–246.

Dworkin, R. (1981a) What is Equality? Part 2: Equality of Resources, *Philosophy and Public Affairs* 10, 283–345.

Easterlin R. (1974) Does Economic Growth Improve the Human Lot? in David, P. A. and Reder, M. W. (eds) *Nations and Households in Economic Growth: Essays in Honor of Moses Abramovitz.* New York: Academic Press.

Easton, M. (2006) The Survival of the Happiest, *New Statesman*, April 24.

Eckersley, R. (2004) *The Green State: Rethinking Democracy and Sovereignty.* MIT Press, Cambridge, MA.

Eckersley, R. (2006) Is modern Western culture a health hazard? *International Journal of Epidemiology*, 35: 252–258.

Edwards, C. and Imrie, R. (2008) Disability and the Implications of the Wellbeing Agenda: Some Reflections from the United Kingdom, *Journal of Social Policy*, 37(3): 337–355.

Elster, J. (1983) *Sour Grapes: Studies in the Subversion of Rationality*. Cambridge: Cambridge University Press.

Ereaut, G., and Whiting, R. (2008) *What do we mean by 'wellbeing'? and why might it matter?* Linguistic Landscapes Research Report no. DCSF-RW073 for the Department of Children, Schools and Families.

Erikson, R. (1993) Descriptions of Inequality: The Swedish Approach to Welfare Research, in Nussbaum and Sen *Quality of Life* Oxford: Oxford University Press.

Erikson, R. and Uusitalo, H. (1987) The Scandinavian Approach to Welfare Research, in Erikson, R., Hansen, E. J., Ringen, S. and Uusitalo, H. (eds) *The Scandinavian Model: Welfare States and Welfare Research*. Armonck: M. E. Sharpe.

Etzioni, A. (2001) The Third Way to a Good Society. Sociale wetenschappen (44e jaargang 2001 nummer 3) 5 http://amitaietzioni.org/A289.pdf (accessed 4/6/2008).

Etzioni, A. (2007) Community Deficit. *Journal of Common Market Studies*, 45(1): 23–42.

Evans, B. and Theobald, K. (2003) Local Agenda 21 and the Shift to 'Soft Governance', in Buckingham, S. and Theobald, K. *Local Environmental Sustainability*. Cambridge: Woodhead Publishing.

Evans, B., Percy, S., and Theobald, K. (2003) *Mainstreaming Sustainability into Local Government Policymaking*. Economic and Social Research Council Report R000223718, UK.

Fairclough, N. (2000) *New Labour, New Language?* London: Routledge.

Fairclough, N. (2003) *Analysing Discourse: textual analysis for social research*, New York: Routledge.

Field, J. (2008) *Social Capital.* Abingdon: Routledge.

Flavin, P., Pacek, A. C. and Radcliff, B. (2011) State Intervention and Subjective Well-Being in Advanced Industrial Democracies, *Politics and Policy*, 39(2): 251–269.

Fleuret, S. and Atkinson, S. (2007) Wellbeing, Health and Geography: a critical review and research agenda, *NZ Geographer*, 63: 106–118.

Flint, J. (2002) Return of the Governors: Citizenship and the New Governance of Neighbourhood Disorder in the UK, *Citizenship Studies*, 6(3), 245–264.

Flyvbjerg, B. (1998) *Rationality and Power: Democracy in practice*. Chicago: University of Chicago Press.

Flyvbjerg, B. (2001) *Making Social Science Matter; Why Social Inquiry Fails and How it Can Succeed Again.* Cambridge: Cambridge University Press.

Forgeard, M., Jayawickreme, E., Kern, M. and Seligman, M. (2011) Doing the Right Thing: Measuring wellbeing for public policy, *International Journal of Wellbeing*, 1(1): 79–106.

Foucault, M. (1991a) Governmentality, in Burchell, G., Gordon, C. and Miller, P. *The Foucault Effect: Studies in Governmentality*. London: Harvester.

Foucault, M. (1991b) (ed. Gordon, C.) *Power/Knowledge: Selected Interviews and Other Writings. 1972–1977*. Brighton: Harvester.

Fraser, H. (2005) Four Different Approaches to Community Participation, *Community Development Journal*, 40(3): 286–300.

Fremeaux, I. (2005) New Labour's Appropriation of the Concept of Community: a Critique, *Community Development Journal*, 40(3): 254–274.

Frey, B.S. and Stutzer, A. (2002). *Happiness and Economics: How the economy and institutions affect human well-being*. Princeton, NJ: Princeton University Press.

Furedi, F. (2008) Pursuit of happiness is personal. *The Australian*, 7 August. http://www.frankfuredi.com/index.php/site/article/233/ (last accessed 18/02/2012).

Furedi, F. (2010) Cameron's happiness Index: counting smiley faces. *Spiked*, 16 November http://www.frankfuredi.com/index.php/site/article/424/ (last accessed 18/02/2012).

Gahin, R., Veleva, V. and Hart, M. (2003) Do Indicators Help Create Sustainable Communities? *Local Environment*, 8(6): 661–666.

Galloway, S. (2005) Literature Review in *Quality of Life and Well-being: Measuring the Benefits of Culture and Sport: Literature Review and Thinkpiece*, Glasgow: Scottish Executive Social Research.

Game, C. (2009) Councillors – an endangered species? *C'llr*, November: 20–21 (Newsletter of the Local Government Information Unit).

Ganesh, S. and McAllum, K. (2010) Well-Being as Discourse: Potentials and Problems for Studies of Organizing and Health Inequalities, *Management Communication Quarterly*, 24(3), 491–498.

Gasper, D. (2007a) Human Well-being: Concepts and Conceptualisations, in McGillivray, M. *Human Well-being: Concept and Measurement.* Basingstoke: Palgrave MacMillan.

Gasper, D. (2007b) Uncounted or Illusory Blessings? Competing Responses to the Easterlin, Easterbrook and Schwartz Paradoxes of Well-being, *Journal of International Development*, 19: 473–492.

Gasper, D. (2009) *Understanding the Diversity of Conceptions of Well-Being and Quality-of Life*, Institute of Social Studies Working Paper 483, The Hague: Institute of Social Studies.

Gasteyer, S. and Flora, C. B. (1999) *Social Indicators: An Annotated Bibliography on Trends, Sources and Development, 1960–1998.* North Central Regional Center for Rural Development, Environmental Protection Agency and Natural Resources Conservation Service Social Sciences Institute. http://www.ncrcrd.iastate.edu/indicators/Indicators1. PDF (accessed 14/2/08).

Geddes, M. (1996) *Extending Democratic Practice in Local Government: A Report to the Commission for Local Democracy.* London: CLD.

Geddes, M., Davies, J. and Fuller, C. (2007) Evaluating Local Strategic Partnerships: Theory and Practice of Change, *Local Government Studies*, 33(1): 97–116.

Giddens, A. (1998) *The Third Way: The Renewal of Social Democracy.* Cambridge: Polity Press.

Goldstein, H. and Leckie, G. (2008) School league tables: what can they really tell us? *Significance*, 5 (2): 67–69.

Gough, I. (2003) *Lists and Thresholds: Comparing the Doyal–Gough Theory of Human Need with Nussbaum's Capabilities Approach*, WeD Working Paper 01. University of Bath: ESRC Research Group on Wellbeing in Developing Countries.

Grenier, P. and Wright, K. (2006) Social Capital in Britain: Exploring the Hall Paradox, *Policy Studies*, 27(1): 27–53.

Hajer, M. (2003) Policy without polity? Policy analysis and the institutional void, *Policy Sciences*, 36:175–195.

Hall, P. (1999) Social Capital in Britain, *British Journal of Political Science*, 29: 417–461.

Hardi, P. and Zdan, T. (1997) *Assessing Sustainability Development: Principles in Practice.* International Institute for Sustainable Development, Canada.

Haugaard, M. (ed.) (2002) *Power: A Reader.* Manchester: Manchester University Press.

Healey, P. (2006) *Collaborative Planning: Shaping Places in Fragmented Societies.* Basingstoke: Palgrave Macmillan.

Healy, P. (2001) Sustainability, Cities and Complexity: re-thinking key dimensions of strategic spatial planning. Conference Paper, SUSPLAN 2001: The Transformation of Sustainable Planning Conference, Newcastle UK.

Heelas, P. and Morris, P. (eds) (1992) *The Values of the Enterprise Culture: The Moral Debate.* London: Routledge.

Helliwell, J. F. and Putnam, R. D. (2004) The social context of well-being, *Philosophical Transactions of the Royal Society London B.* 359: 1435–1446.

Higginson, S., Sommer, F. and Terry, A. (2003) *Making Indicators Count: Using Quality of Life Indicators in Local Governance – Identifying the Missing Link.* New Economics Foundation and University of West England.

Hoggett, P. (1997) *Contested Communities: Experiences, Struggles, Policies.* Bristol: Policy Press.

Hothi, M., Bacon, N., Brophy, M., and Mulgan, G. (2008) *Neighbourliness + Empowerment = Wellbeing: Is there a formula for happy communities?* London: Young Foundation.

Howarth, D. (2000) *Discourse.* Buckinghamshire: Open University Press.

Howarth, J. and Hart, G. (eds) (2007) *Well-Being: Individual, Community and Social Perspectives.* Basingstoke: Palgrave Macmillan.

Huppert, F., Baylis, N. and Keverne, B. (eds) (2005) *The science of well-being*, Oxford: Oxford University Press.

Huppert, F. A., Marks, N., Clark, A., Siegrist, J., Stutzer, A., Vittersø, J. and Wahrendorf, M. (2009) Measuring Well-being Across Europe: Description of the ESS Well-being Module and Preliminary Findings, *Social Indicators Research*, 91: 301–315.

Innes, J. E., (1990) *Knowledge and Public Policy: The Search for Meaningful Indicators*, Transaction Publishers, New Jersey.

Jackson, T. (2005) *Motivating Sustainable Consumption: a review of evidence on consumer behaviour and behaviour change.* SDRN Research Review http://sdrnadmin.rechord. com/wp-content/uploads/motivatingscfinal_000.pdf.

Jackson, T. (2006) Consuming Paradise – towards a social and cultural psychology of sustainable consumption, in Jackson, T. (ed.) *The Earthscan Reader in Sustainable Consumption.* London: Earthscan.

Jackson, T. (2009) *Prosperity Without Growth: Economics for a Finite Planet.* London: Earthscan.

James, O. (2005) *Affluenza: How to Be Successful and Stay Sane.* London: Vermilion.

John Thompson and Partners (2007) *Bates Colliery Strategic Development Guide.* Report prepared for South East Northumberland North Tyneside Regeneration Initiative, English Partnerships and Blyth Valley Borough Council by Thompson & Partners. Project code BVC.BCM. London, UK .

Johns, H. and Ormerod, P. (2007) *Happiness, Economics and Public Policy.* London: Institute of Economic Affairs.

Johnson, L. B. (1964) Great Society, Speech delivered at Michigan University. May 22. http://www.lbjlib.utexas.edu/johnson/lbjforkids/gsociety_read.shtm (accessed 15/9/2011).

Jones, A. (2008) *The New Geography of Wealth and Poverty*, Local Government Information Unit and STEER, Policy Briefing PB 1719/08L http://www.lgiu.gov.uk/ briefing-detail.jsp?id=1719&md=0&section=briefing (accessed 5/2/08).

Jordan, B. (2008)*Welfare and Well-being; social value in public policy*, Bristol: Policy Press.

Jupp, E. (2007) Participation, local knowledge and empowerment: researching public space with young people, *Environment and Planning A* 39, 2832–2844.

Kasser, T. and Ryan, R. M. (1996) Further examining the American dream: Differential correlates of intrinsic and extrinsic goals, *Personality and Social Psychology Bulletin*, 22: 280–287.

Keough, N. (2005) The Sustainable Calgary Story: A Local Response to a Global Challenge, in Phillips, R. (ed.) *Community Indicators Measuring Systems*. Aldershot and Burlington: Ashgate.

Kingdon, J. W. (1995) *Agendas, Alternatives, and Public Policies*. Second edition. New York: Harper Collins.

Kjaer, A. M. (2004) *Governance*. Cambridge: Polity Press.

Kothari, U. (2001) Participatory Development: Power, Knowledge and Social Control, in Kothari, U. & Cooke, B. (eds) *Participation: the New Tyranny?* London: Zed Books.

Laessoe, J. (2007) Participation and Sustainable Development: The Post-ecologist Transformation of Citizen Involvement in Denmark, *Environmental Politics*, 16(2), 231–250.

Lash, S., Szerszynski, B. and Wynne, B. (eds) (1996) *Risk, Environment and Modernity: Towards a New Ecology*. Sage, London.

Lawrence, J. G. (1998) Getting the Future That You Want: The Role of Sustainability Indicators, in Warburton, D. (ed.) *Community and Sustainable Development: Participation in the Future*. London: Earthscan.

Layard, R. (2003) Happiness: Has Social Science a Clue? Lionel Robbins Memorial Lecture, 2 March 2003, LSE.

Layard, R. (2006) *Happiness: lessons from a new science*. London: Allen Lane.

Levett, R. (1998) Sustainability Indicators – Integrating Quality of Life and Environmental Protection, *Journal of the Royal Statistical Society, Series A (Statistics in Society)*, 161(3), 291–302.

Levett-Therivel Sustainability Consultants (2007) *Wellbeing: International Policy Interventions*. Report to Defra.

LGIU (2005) *Local Strategic Partnerships and Sustainability*. Policy Briefing PB030/05, Local Government Information Unit.

Lingayah, S. and Sommer, F. (2001) *Communities Count: The LITMUS Test: Reflecting Community Indicators in the London Borough of Southwark*. New Economics Foundation and London Borough of Southwark, London.

Llewellyn Davies (2005) *Blyth Estuary Framework Plan*. Report prepared for SENNTRi, January.

Local Government Association and Audit Commission (2002). *Quality of life indicators. A survey of pilots and non-pilots*. Research Briefing 18.

Lowndes, V. (1999) Management Change in Local Governance, in Stoker, G. *The New Management of British Local Governance*, London: MacMillan.

Luke, T. W. (1995) On Environmentailty: Geo-Power and Eco-Knowledge in the Discourses of Contemporary Environmentalism, *Cultural Critique*, Fall, 57–81.

Lukes, S. (2005) *Power: A Radical View* (Second Edition). Palgrave Macmillan, Basingstoke.

MacGillivray, A. (1998) Turning the Sustainability Corner: How to Indicate Right, in Warburton, D. (ed.) (1998) *Community and Sustainable Development: Participation in the Future*. London: Earthscan.

MacGillivray, A., Weston, C. and Unsworth, C. (1998) *Communities Count! A Step by Step Guide to Community Sustainability Indicators*, New Economics Foundation, London.

Macnaghten P. and Urry J. (1998) *Contested Natures*. London, Sage.

Maloney, W., Smith, G., and Stoker, G. (2000) Social Capital and Urban Governance: Adding a More Contextualized 'Top-down' Perspective, *Political Studies*, 48: 802–820.

Manderson, L. (2005) *Rethinking Wellbeing*: Perth, Australia: API Network.

Marinetto, M. (2003) Who Wants to be an Active Citizen? The Politics and Practice of Community Involvement, *Sociology*, 37(1): 103–120.

Marks, N., Shah, H. and Westall, A. (2004) *The power and potential of well-being indicators*. London: New Economics Foundation.

Marks, N., Abdallah, S., Simms, A. and Thompson, S. (2006a) *The (un)Happy Planet Index*. London: New Economics Foundation.

Marks, N., Thompson, S., Eckersley, R., Jackson, T., and Kasser, T. (2006b) *Sustainable development and well-being; relationships, challenges and policy implications*. DEFRA Project 3b. New Economics Foundation, London.

Marmot, M., Allen, J., Goldblatt, P., Boyce, T., McNeish, D., Grady, M. and Geddes, I. (2010). *The Marmot Review: Fair Society, Healthy Lives. Strategic Review of Health Inequalities in England post-2010*. London: The Marmot Review.

Marsh, D. and Stoker, G. (eds) (2002) *Theory and Methods in Political Science*. Palgrave Macmillan, Hampshire.

Marvin, S. and Guy, S. (1997) Constructing Myths Rather than Sustainability: The Transition Fallacies of the New Localism, *Local Environment*, 2(3): 311–318.

Maslow, A. (1943) A theory of human motivation, *Psychological Review*, 50: 370–396.

Maslow, A. (1954) *Motivation and Personality*. New York: Harper.

Max-Neef, M. (2011) The death and rebirth of economics. Preface, in Rauschmayer, F., Omann, I. and Frühmann, J. (2011) *Sustainable Development: Capabilities, Needs and Well-being*. Abingdon: Routledge.

McAllister, F. (2005) *Wellbeing Concepts and Challenges*, discussion paper prepared for the Sustainable Development Research Network. http://www.sd-research.org.uk/post.php?p=128 (last accessed 18 02 2012).

McAlpine, P. and Birnie, A. (2005) Is There a Correct Way of Establishing Sustainability Indicators? The Case of Sustainability Indicator Development on the Island of Guernsey, *Local Environment*, 10(3), 243–257.

McGillivray, M. (2007a) *Human Well-being: Concept and Measurement*. Palgrave MacMillan, Basingstoke.

McGillivray, M. (2007b) Human Well-being: Issues, Concepts and Measures, in McGillivray, M. *Human Well-being: Concept and Measurement*. Palgrave MacMillan, Basingstoke.

Meadows, D. H., Meadows, D. L., Randers, J. and Behrens, W. W. (1972) *The Limits to Growth*. New York: Universe Books.

Mebratu, D. (1998) Sustainability and sustainable development: historical and conceptual view, *Environmental Impact Assessment Review*, 18 493–520.

Mguni, N. and Bacon, N. (2010) *Taking the temperature of local communities: the wellbeing and Resilience Measure (WARM)*, London: Young Foundation.

Michalos, A. (1997) Combining Social, Economic and Environmental Indicators to Measure Sustainable Human Wellbeing, *Social Indicators Research*, 40(1–2): 221–258.

Michalos, A. C. (1999) Reflections on Twenty-Five Years of Quality-of-Life Research, *Feminist Economics*, 5(2): 119–123.

Michalos, A. C. (2011) What did Stiglitz, Sen and Fitoussi Get Right and What Did They Get Wrong? *Social Indicators Research*, 102: 117–129.

Michalos, A. C., Smale, B., Labonté, R., Muhajarine, N., Scott, K., Moore, K., Swystun, L., Holden, B., Berardin, H., Dunning, B., Graham, P., Guhn, M., Gadermann, A. M., Zumbo, B. D., Morgan, A., Brooker, A.-S. and Hyman, I. (2011) *The Canadian Index of Wellbeing*. Technical Report 1.0, Waterloo, Ontario: Canadian Index of Wellbeing and University of Waterloo.

Miller, W., Dickson, M. and Stoker, G. (2000) *Models of Local Governance: Public Opinion and Political Theory in Britain*. Basingstoke: Palgrave.

Mulgan, G. (1997) *Connexity: How to Live in a Connected World*. London: Chatto and Windus.

Myerson, G. and Rydin, Y. (1996a) *The Language of the Environment*. London: UCL Press.

Myerson, G. and Rydin, Y. (1996b) Sustainable Development: The Implications of the Global Debate for the Land Use Planning System, in Buckingham-Hatfield, S. and Evans, B. (eds) *Environmental Planning and Sustainability*. Chichester: Wiley.

Naess, A. (1989) *Ecology, Community and Lifestyle*. Cambridge: Cambridge University Press.

Najam, A. (2005) Developing Countries and Global Environmental Governance: From Contestation to Participation to Engagement, *International Environmental Agreements*, 5: 303–321.

Nettle, D. (2006) *Happiness: the science behind your smile*. Oxford: Oxford University Press.

Neumayer, E. (2007) Sustainability and Well-being Indicators, in McGillivray, M. *Human Well-being: Concept and Measurement*. Basingstoke: Palgrave MacMillan.

NEF, New Economics Foundation (2004) *Chasing Progress: Beyond Measuring Economic Growth* The Power of Well-being 1 – Research Paper Summary http://www.neweconomics.org/gen/uploads/izgu3e45trldy0e2qzb4wt5516032004125132.pdf (accessed 7/12/2004).

Nissel, M. (1995) Social Trends and Social Change, *Journal of the Royal Statistical Society A* 158(3), 491–504.

Noll, H. (2000) *Social Indicators and Social Reporting: the International Experience*. http://www.ccsd.ca/noll1.html.

Norman, J. (2010) *The Big Society: The Anatomy of the New Politics*. Buckingham: The University of Buckingham Press.

Northumberland InfoNet (2005) *State of Northumberland 2005*. Report prepared for Northumberland Strategic Partnership. http://www2.northumberland.gov.uk/reports/CPA_JAR_CA/cpa/CD1-Community-Strategy/StateNLand05.pdf (accessed July 2008).

Nussbaum, M. (1998) *Sex and Social Justice*. New York: Oxford University Press.

Nussbaum, M. (1999) *Sex and Social Justice*. Oxford: Oxford University Press.

Nussbaum, M. (2000) *Women and Human Development*. New York: Cambridge University Press.

Nussbaum, M. (2003) Capabilities as Fundamental Entitlements: Sen and Social Justice, *Feminist Economics*, 9(2–3), 33–59.

Nussbaum, M. (2005) Wellbeing, Contracts and Capabilities, in Manderson, L. (2005) *Rethinking Wellbeing*. Perth, Australia: API Network.

Nussbaum, M. and Sen, A. (eds) (1993) *The Quality of Life*. New York: Oxford University Press.

ODPM (2006) *Promoting Effective Citizenship and Community Empowerment: A guide for local authorities on enhancing capacity for public participation*. ODPM, London & Centre for Local and Regional Government Research, Cardiff University.

Offer, A. (2000) *Economic Welfare Measurements and Human Well-being*, Discussion Papers in Social and Economic History number 34, January, University of Oxford.

O'Neill, J. (1993) *Ecology, Policy and Politics: Human Well-being and the Natural World*. London: Routledge.

O'Neill, J. (2011) The Overshadowing of Needs, in Rauschmayer, F., Omann, I. and Frühmann, J. (eds) *Sustainable Development: Capabilities, needs and well-being*. Abingdon: Routledge.

ONS, Office of National Statistics (2010) *National Statistician launches well-being debate*. Press release, 25 November. ONS: Newport

200 References and further reading

ONS, Office of National Statistics (2011) *Overview of Societal Wellbeing.* http://www. statistics.gov.uk/hub/people-places/communities/societal-wellbeing/index.html.

Opschoor, J. B. and Weterings, R. (1994) Environmental utilisation space: an introduction. *Milieu- Tijdschrift voor Milieukunde*, 9(5): 198–205.

O'Riordan, T. (1988) The Politics of Sustainability, in Turner R.K. (ed.) *Sustainable Environmental Management: Principles and Practice.* London: Belhaven Press.

O'Riordan, T. (ed.) (2001) *Globalism, Localism and Identity: Fresh Perspectives on the Transition to Sustainability.* London: Earthscan.

Ormerod, P. (2007) Against Happiness, *Prospect Magazine*, 133: 1–6.

Ortega-Cerdà, M. (2005) Sustainability Indicators as Discursive Elements. Paper submitted at 6th International Conference of the European Society for Ecological Economics in Lisbon 16/6/2005. http://ecoman.dcea.fct.unl.pt/projects/esee2005/papers/138_1104503992647_fullpaper.pdf (accessed 3/7/2007).

Oughton, E. and Wheelock, J. (2006) The relationship between consumption and production: conceptualizing well-being inside the household, in Clary J., Dolfsma, W. and Figart, D. (2006) *Ethics and the Market: Insights from Social Economics.* Abingdon: Routledge.

Oughton, E., Wheelock, J. and Baines, S. (2003) Micro-businesses and Social Inclusion in Rural Households: A Comparative Analysis, *Sociologica Ruralis*, 43(4): 331–348.

Pain, R. and Kindon, S. (2007) Participatory Geographies, *Environment and Planning A*, 39: 2807–2812.

PASTILLE (2002) *Indicators into Action. Local Sustainability Indicator Sets in Their Context.* European Union FP5. London: London School of Economics.

Pateman, C. (1970) *Participation and Democratic Theory*, London: Cambridge University Press.

Pattie, C., Seyd, P. and Whitely, P. (2003) Citizenship and Civic Engagement: Attitudes and Behaviour in Britain, *Political Studies*, 51(4) 616–33.

Peck, J. (2010) Zombie neoliberalism and the ambidextrous state. *Theoretical Criminology*, 14(1): 104–110.

Peck, J., Theodore, N. and Brenner, N. (2009) Postneoliberalism and its Malcontents, *Antipode*, 41 (S1 2009): 94–116.

Peterson, A. and Lupton, D. (2000) *The New Public Health: Health and Self in the Age of Risk.* London: Sage.

Phillips, R. (ed.) (2005) *Community Indicators Measuring Systems.* Aldershot and Burlington: Ashgate.

Phillips, D. (2006) *Quality of Life: Concept, Policy and Practice.* London: Routledge.

Power, M. (1994) *The Audit Explosion.* London: Demos.

Prilleltensky, I. and Prilleltensky, O. (2007) Webs of Well-Being: the Interdependence of Personal, Relational, Organizational and Communal Well-Being, in Howarth, J. and Hart, G. (eds) *Well-Being: Individual, Community and Social Perspectives.* Basingstoke: Palgrave Macmillan.

Putnam, R. D. (2000) *Bowling Alone: The Collapse and Revival of American Community.* New York: Simon and Schuster.

Raco, M. (2005) Sustainable Development, rolled-out Neoliberalism and Sustainable Communities, *Antipode*, 37: 324–346.

Raco, M. (2007) Securing Sustainable Communities; Citizenship, Safety and Sustainability in the New Urban Planning, *European Urban and Regional Studies*, 14(1): 305–320.

Rallings, C. and Thrasher, M. (2006) 'Just Another Expensive Talking Shop': Public Attitudes and the 2004 Regional Assembly Referendum in the North East of England, *Regional Studies*, 40(8): 927–936.

Randall, N. (2009) No Friends in the North? The Conservative Party in Northern England, *The Political Quarterly*, 80(2): 184–192.

Rapley, M. (2003) *Quality of Life Research: A Critical Introduction*. London: Sage.

Rauschmayer, F., Omann, I. and Frühmann, J. (2011) *Sustainable Development: Capabilities, needs and well-being*. Routledge: Abingdon.

Rawls, J. (1971) *A Theory of Justice*. Cambridge, MA: Harvard University Press.

Reboratti, C. E. (1999) Territory, scale and sustainable development, in Becker, E. and Jahn, T. (eds) *Sustainability and the Social Sciences: A Cross-disciplinary Approach to Integrating Environmental Considerations into Theoretical Reorientation*. London: Zed Books.

Rey-Valette, H., Laloe, F. and Le Fur, J. (2007) Introduction to the key issue concerning the use of sustainable development indicators, *International Journal of Sustainable Development*, 10(1–2): 4–13.

Rhodes, R. A. W. (1997) *Understanding Governance: Policy Networks, Governance, Reflexivity and Accountability*. Buckingham: Open University Press.

Robeyns, I. (2003) Sen's Capability Approach and Gender Inequality: Selecting Relevant Capabilities. *Feminist Economics*, 9 (2–3): 61–92.

Robeyns, I. (2005) The Capability Approach: a theoretical survey, *Journal of Human Development*, 6(1) 93–110.

Robeyns, I. (2008) Sen's Capability Approach and Feminist Concerns, in Alkire, S., Qizilbash, M. and Comim, F. (eds) *The Capability Approach: Concepts, Measures and Applications*. Cambridge: Cambridge University Press.

Rose, N. (1992) Governing the Enterprising Self, in Heelas, P. and Morris, P. (eds) *The Values of the Enterprise Culture: The Moral Debate*. London: Routledge.

Rose, N. (1996) *Inventing Our Selves. Cambridge:* Cambridge University Press.

Russel, D. (2007) The United Kingdom's Sustainable Development Strategies: Leading the Way or Flattering to Deceive? *European Environment*, 17: 189–200.

Rydin, Y. (2003) *Conflict, Consensus, and Rationality in Environmental Planning: An Institutional Discourse Approach*. Oxford: Oxford University Press.

Rydin, Y. (2007) Indicators as a governmental technology? The lessons of community-based sustainability indicator projects, *Environment and Planning D: Society and Space*, 25: 610–624.

Ryan, R. M. and Deci, E. L. (2000) Self-determination theory and the facilitation of intrinsic motivation, social development and well-being, *American Psychologist*, 55: 68–78.

Rydin, Y. and Pennington, M. (2000) Public Participation and Local Environmental Planning: the collective action problem and the potential of social capital, *Local Environment*, 5(2): 153–169.

Rydin, Y., Holman, N. and Wolff, E. (2003) Local Sustainability Indicators, *Local Environment*, 8(6): 581–589.

Ryff, C. (1989) Happiness is everything, or is it? Explorations on the meaning of psychological well-being, *Journal of Personality and Social Psychology*, 57: 1069–1081.

Ryff, C. and Keyes, C. (1995) The structure of psychological well-being revisited, *Journal of Personality and Social Psychology*, 69: 719–727.

Schalock, R. L. and Verdugo, M. A. (2002) *Quality of life handbook for human service practitioners*, Washington DC: American Association on Mental Retardation.

Schmuck, P., Sheldon, K. M., Ryan, R. M., Deci, E. L., and Kasser, T. (2004). The independent effects of goal contents and motives on well-being: It's both what you pursue and why you pursue it, *Personality and Social Psychology Bulletin*, 30: 475–486.

Schoch, R. (2006) *The Secrets of Happiness: Three Thousand Years of Searching for The Good Life*. London: Profile Books.

Scott, K. (2009) Measuring local well-being: towards sustainability?: A case study of developing quality of life indicators at Blyth Valley Borough Council, PhD Thesis, University of Newcastle Upon Tyne.

Scott, K. (2012) A 21st Century Sustainable Community: Discourses of Local Wellbeing, in Atkinson, S., Fuller, S. and Painter, J. (eds) *Wellbeing and Place*. Aldershot: Ashgate.

Seedhouse, D. (2001) *Health: The foundations for achievement*. Chichester, UK: John Wiley.

Seligman, M. (2011). *Flourish*. New York: Simon & Schuster.

Sen, A. (1980) Equality of What? in McMurrin, S. (ed.)*Tanner Lectures on Human Values*. Cambridge: Cambridge University Press.

Sen, A. (1993) Capability and Wellbeing, in Nussbaum, M. and Sen, A. (eds) *The Quality of Life*. New York: Oxford University Press.

SENNTRi, South East Northumberland North Tyneside Regeneration Initiative (SENNTRi) (2006) *Bates Colliery Proposed Residential Site Strategic Development Guide*. Blyth, Newcastle.

Shah, H. and Marks, N. (2004) *A well-being manifesto for a flourishing society*. London: New Economics Foundation.

Shaw, K. and Robinson, F. (2007) 'The End of the Beginning'? Taking Forward Local Democratic Renewal in the Post-Referendum North East, *Local Economy*, 22(3): 243–260.

Shepherd. S. (2007) *Quality of Life Indicators: Factors which Affect their Utility to Decision Makers in the Public Sector*. PhD Thesis, University of the West of England, Bristol.

Sirgy, M. J., Rahtz, D. and Lee, D. (eds) (2004) *Community Quality of Life Indicators: Best Cases*, Dordecht: Kluwer Academic Publishers.

Sirgy, M. J., Michalos, C., Ferriss, A. L., Easterlin, R., Patrick, D., and Pavot, W. (2006) The Quality of Life (QOL) Research Movement: Past, Present, and Future, *Social Indicators Research*, 76(3), 343–466.

Sirgy, M. J., Phillips, R. and Rahtz, D. (eds) (2011) *Community Quality-of-Life Indicators – Best Cases V*, Dordrecht: Springer.

Skidmore, P., Bound, K. and Lownsbrough, H. (2006) *Community Participation: Who Benefits?* York: Joseph Rowntree Foundation.

Sommer, F. (2000) Monitoring and Evaluating Outcomes of Community Involvement – The LITMUS Experience, *Local Environment*, 5(4): 483–491.

Steuer, N. and Marks, N. (2008) *Local Wellbeing: Can We Measure It?* London: New Economics Foundation.

Stewart, J. (2000) *The Nature of British Local Government*. MacMillan, Basingstoke.

Stiglitz, J. E., Sen, A. and Fitoussi, J. P. (2009) *Report by the Commission on the Measurement of Economic Performance and Social Progress*. http://www.stigliz-sen-fitoussi.fr/ (accessed 14/3/2010).

Stoker, G. (ed.) (1999) *The New Management of British Local Governance*. Basingstoke: MacMillan.

Stoker, G. (2000) Introduction, in Stoker, G. (ed.) *The New Politics of British Local Government*, London: Palgrave Macmillan.

Stutzer, A and Frey, B.S. (2006) Political participation and procedural utility: An empirical study, *European Journal of Political Research*, 45: 391–418.

Sullivan, H. (2005) Is enabling enough? Tensions and dilemmas in New Labour's strategies for joining up local governance, *Public Policy and Administration*, 20(4): 10–24.

Taylor, P. (2005) Surge in deaths spark drug fear. *Evening Chronicle*, Newcastle, England. October 29 2005.

Taylor, M. (2007) Community Participation in the Real World: Opportunities and Pitfalls in New Governance Spaces, *Urban Studies*, 44(2): 297–317.

Tenenbaum, E. and Wildavsky, A. (1984) Why Policies Control Data and Data Cannot Determine Policies, *Scandinavian Journal of Management Studies*, 1(2): 83–100.

Terry, A. (2008) Community sustainable-development indicators: a useful participatory technique or another dead end? *Development in Practice*, 18(2): 223–234.

Thompson, K. (1992) Individual and Community in Religious Critiques of the Enterprise Culture, in Heelas, P. and Morris, P. (eds) *The Values of the Enterprise Culture: The Moral Debate*. London: Routledge.

Tickell, A., John, P. and Musson, S. (2005) The North East Region Referendum Campaign of 2004: Issues and Turning Points, *Political Quarterly*, 76(4): 488–496.

Treanor, P. (1996) *Why Sustainability is Wrong*. http://www.inter.nl.net/users/paul.Treanor/ sustainability.htm (accessed 20/4/1998, website no longer accessible).

Tsai, M.-C. (2011) If GDP is Not the Answer, What is the Question? The Juncture of Capabilities, Institutions and Measurement in the Stiglitz–Sen–Fitoussi Report, *Social Indicators Research*, 102: 363–372.

Tunstall, R. and Lupton, R. (2010) *Mixed Communities: Evidence Review*. London: Department for Communities and Local Government.

Turnhout, E., Hisschemoller, M. and Eijsackers, H. (2007) Ecological Indicators: Between the two fires of science and policy, *Ecological Indicators*, 7, 215–228.

Tyler, T.R. (2000) Social justice: Outcome and procedure, *International Journal of Psychology*, 35, 117–125

UK Government (1999) Modernising Government White Paper, The Stationary Office, London. http://www.archive.official-documents.co.uk/document/cm43/4310/4310.htm.

UK Government (2000) Local Government Act 2000 Part 1. HMSO.

UK Government Cabinet Office (2008) *Realising Britain's Potential: Future Strategic Challenges for Britain*, Discussion Paper. The Strategy Unit, Cabinet Office. http:// www.cabinetoffice.gov.uk/upload/assets/www.cabinetoffice.gov.uk/strategy/strategic_ challenges.pdf (accessed 13/2/2008).

UK Government Office of Public Sector Information (2007) Sustainable Communities Act 2007. http://www.opsi.gov.uk/acts/acts2007/ukpga_20070023_en_1#Legislation-Preamble (accessed 12/1/08).

United Nations Conference on Environment and Development, UNCED (1992) *Agenda 21*. UNCED.

United Nations Economic Commission for Europe (UNECE)(1998) Convention on Access to Information, Public Participation in Decision-Making and Access to Justice in Environmental Matters. http://www.unece.org/env/pp/documents/cep43e.pdf.

Veenhoven, R. (1996) Happy Life-expectancy: a comprehensive measure of quality-of-life in nations. *Social Indicators Research*, 39: 1–58.

Vlek, C., Skolnik, M. and Gatersleben, B. (1998) Sustainable development and quality of life: expected effects of prospected changes in economic and environmental conditions, *Zeitschrift für Experimentelle Psychologie*, 45(4): 319–333.

Voisey H., Walters, A. and Church, C. (2001) Local Identity and Empowerment in the UK, in O'Riordan, T. (ed.) *Globalism, Localism and Identity: Fresh Perspectives on the Transition to Sustainability*. London: Earthscan.

Waddington, D., Critcher, C., Dicks, B. and Parry, D. (2001) *Out of the Ashes?: The Social Impact of Industrial Contraction and Regeneration on Britain's Mining Communities*. London: Jessica Kingsley.

Wagle, U. (2000) The policy science of democracy: The issues of methodology and citizen participation, *Policy Sciences*, 33: 201–223.

Warburton, D. (1998a) Participation in Conservation: grasping the nettle, *Ecos*, 19(2).

Warburton, D. (ed.) (1998b) *Community and Sustainable Development: Participation in the Future.* London: Earthscan.

White, S. and Pettit, J. (2007) Participatory Approaches and the Measurement of Human Well-being, in McGillivray, M. *Human Well-being: Concept and Measurement.* Basingstoke: Palgrave MacMillan.

Wilkinson, R. G. and Pickett, K. E. (2006) Income Inequality and Population Health: A Review and Explanation of the Evidence, *Social Science and Medicine*, 62(7): 1768–1784.

Wilkinson, R. G. and Pickett, K. E. (2010) *The Spirit Level: Why equality is better for everyone*, London: Penguin.

Williams, A., Holden, B., Krebs, P., Muhajarine, N., Waygood, K., Randall, J. and Spence, C. (2008) Knowledge translation strategies in a community–university partnership: examining local Quality of Life (QoL), *Social Indicator Research*, 85: 111–125.

Woodward, R. (1996) 'Deprivation' and 'the Rural': an Investigation into Contradictory Discourses. *Journal of Rural Studies*, 12 (1): 55–67.

World Commission on Environment and Development (WCED) (1987) *Our Common Future.* Oxford: Oxford University Press.

Wright, S. F., Parry, J., Mathers, J., Jones, S. and Orford, J. (2006) Assessing the Participatory Potential of Britain's New Deal for Communities: Opportunities for and constraints to 'bottom-up community participation', *Policy Studies*, 27(4): 347–361.

Zidanšek, A. (2007) Sustainable development and happiness in nations, *Energy*, 32(6): 891–897.

Zittel, T. and Fuchs, D. (2007) *Participatory Democracy and Political Participation.* London: Routledge.

Zittoun, P. (2006) Indicateurs et cartographie dynamique du bruit, un instrument de reconfiguration des politiques publiques?. Dossier 8: Méthodologies et pratiques territoriales de l'évaluation en matière de développement durable. *Développement Durable et Territoires.* http://developpementdurable.revues.org/document3261.html. (accessed 4/4/08).

Index